S0-AYT-818

Cell and Developmental Biology
of the Eye

Cell and Developmental Biology
of the Eye

Series Editors

Joel B. Sheffield and S. Robert Hilfer

Ocular Size and Shape: Regulation During Development

Cellular Communication During Ocular Development

Molecular and Cellular Basis of Visual Acuity

Heredity and Visual Development

The Microenvironment and Vision

The Proceedings of the Philadelphia Symposia on Ocular
and Visual Development

The Microenvironment and Vision

Edited by
Joel B. Sheffield
S. Robert Hilfer

With 60 Figures

Springer-Verlag
New York Berlin Heidelberg
London Paris Tokyo

Joel B. Sheffield
S. Robert Hilfer
Department of Biology
Temple University
Philadelphia, Pennsylvania 19122, U.S.A.

On the cover: New capillary formation. See page 30.

Library of Congress Cataloging-in-Publication Data
The microenvironment and vision.
  (Cell and developmental biology of the eye)
  Based on the 10th Symposium on Ocular and
  Visual Development, held Oct. 1985 in Philadelphia.
  Includes bibliographies and index.
  1. Eye—Congresses.   2. Extracellular matrix—
Congresses.   3. Eye—Growth—Congresses.   4. Eye—
Diseases and defects—Etiology—Congresses.
I. Sheffield, Joel B.   II. Hilfer, S. Robert.
III. Symposium on Ocular and Visual Development
(10th : 1985 : Philadelphia, Pa.)   IV. Series.
[DNLM: 1. Environment—congresses.   2. Environmental
Exposure—congresses.   3. Eye—metabolism—congresses.
4. Eye Diseases—etiology—congresses.   5. Vision—
physiology—congresses. W3 SY5363 10th 1985m /
WW 103 M626 1985]
QP474.M53   1987   612'.84   87-9552

©1833

@ P
474
.M53
1987

©1987 by Springer-Verlag New York Inc.
All rights reserved. This work may not be translated or copied in whole or in part without the written
permission of the publisher (Springer-Verlag, 175 Fifth Avenue, New York, New York 10010, USA),
except for brief excerpts in connection with reviews or scholarly analysis. Use in connection with any
form of information storage and retrieval, electronic adaptation, computer software, or by similar or
dissimilar methodology now known or hereafter developed is forbidden.
The use of general descriptive names, trade names, trademarks, etc. in this publication, even if the
former are not especially identified, is not to be taken as a sign that such names, as understood by
the Trade Marks and Merchandise Marks Act, may accordingly be used freely by anyone.
While the advice and information in this book are believed to be true and accurate at the date of going
to press, neither the authors nor the editors nor the publisher can accept any legal responsibility for
any errors or omissions that may be made. The publisher makes no warranty, express or implied, with
respect to the material contained herein. Permission to photocopy for internal or personal use, or the
internal or personal use of specific clients, is granted by Springer-Verlag, New York Inc. for libraries
and other users registered with the Copyright Clearance Center (CCC), provided that the base fee of
$0.00 per copy, plus $0.20 per page is paid directly to CCC, 21 Congress Street, Salem, MA 01970,
USA. Special requests should be addressed directly to Springer-Verlag New York, 175 Fifth Avenue,
New York, NY 10010 U.S.A. 95640-8/87 $0.00 + .20
Printed and bound by Arcata Graphics/Halliday, West Hanover, Massachusetts.
Printed in the United States of America.

9 8 7 6 5 4 3 2 1

ISBN 0-387-96540-8 Springer-Verlag New York Berlin Heidelberg
ISBN 3-540-96540-8 Springer-Verlag Berlin Heidelberg New York

Series Preface

The eye has fascinated scientists from the earliest days of biological investigation. The diversity of its parts and the precision of their interaction make it a favorite model system for a variety of developmental studies. The eye is a particularly valuable experimental system not only because its tissues provide examples of fundamental processes, but also because it is a prominent and easily accessible structure at very early embryonic ages.

In order to provide an open forum for investigators working on all aspects of ocular development, a series of symposia on ocular and visual development was initiated in 1973. A major objective of the symposia has been to foster communication between the basic research worker and the clinical community. It is our feeling that much can be learned on both sides from this interaction. The idea for an informal meeting allowing maximum exchange of ideas originated with Dr. Leon Candeub, who supplied the necessary driving force that made the series a reality. Each symposium has concentrated on a different aspect of ocular development. Speakers have been selected to approach related topics from different perspectives.

This book series, "Cell and Developmental Biology of the Eye," is derived from the Philadelphia symposia on ocular and visual development. Previous volumes are listed on the series page. We hope that the introduction of this proceedings series will make the results of research on ocular cell and developmental biology more widely known and more easily accessible.

Preface

The 10th Symposium on Ocular and Visual Development was held in October 1985 to explore current knowledge on the molecular composition of the matrix surrounding the cells of the ocular system. The environment surrounding the cornea, lens, and retina is of particular importance for normal function, and many degenerative diseases, such as cataract and retinal degeneration, result from failure of normal metabolic processes. In addition, environmental factors such as light and diet have been demonstrated to influence cellular metabolism and lead to ocular injury. The papers in this volume consider different aspects of the microenvironment and its role in ocular metabolism. These range from the ordered construction of the cornea and the factors stimulating vascularization to the effects of light and oxidative damage and therapeutic use of vitamin E. These subjects are treated from various perspectives, including the generation of free radicals and other damaging oxygen species, measurement of the effects of various agents, and clinical assessment of therapy. This multidisciplinary approach allows the reader to understand the complex nature of the problem and to form his or her own judgments.

We are indebted to our colleagues, Dr. L. Andrews and Dr. P. Dayhaw-Barker of the Pennsylvania College of Optometry, for their expertise and assistance in the organization of the meeting, to the speakers for their presentations and contributions to this volume, and to the reviewers of the manuscripts for their helpful comments.

This symposium could not have been held without the generous support of the Temple University College of Arts and Sciences and the Pennsylvania College of Optometry. We also thank Hoffmann-La Roche, Inc. and the Rae S. Uber Foundation for their support. This volume was prepared with the skillful assistance of Mai Tran, to whom we are indebted.

Joel B. Sheffield
S. Robert Hilfer

Philadelphia, Pennsylvania

Contents

Contributors

Robert E. Anderson, Cullen Eye Institute, Baylor College of Medicine, Houston, Texas 77030, U.S.A.

David E. Birk, Department of Pathology, Robert Wood Johnson Medical School, University of Medicine and Dentistry of New Jersey, Piscataway, New Jersey 08854, U.S.A.

C. D. B. Bridges, Cullen Eye Institute, Department of Ophthalmology, Baylor College of Medicine, Houston, Texas 77030, U.S.A.

Barbara Buckley, Department of Medicine, Duke University Medical Center, Durham, North Carolina 27710, U.S.A.

Ann H. Bunt-Milam, Department of Ophthalmology, University of Washington School of Medicine, Seattle, Washington 98195, U.S.A.

Patricia A. D'Amore, Departments of Surgical Research and Pathology, The Children's Hospital and Harvard Medical School, Boston, Massachusetts 02115, U.S.A.

Neil Finer, Department of Pediatrics, University of Alberta, Edmonton, Alberta, Canada.

Bruce Freeman, Departments of Anesthesiology and Biochemistry, The University of Alabama at Birmingham, Birmingham, Alabama 35294, U.S.A.

Vinod Gaur, Department of Ophthalmology, University of Washington School of Medicine, Seattle, Washington 98195, U.S.A.

R. Kennon Guerry, Medical College of Virginia, Virginia Commonwealth University, Richmond, Virginia 23298, U.S.A.

William T. Ham, Jr., Medical College of Virginia, Virginia Commonwealth University, Richmond, Virginia 23298, U.S.A.

Kimiko Hayashi, Department of Pathology, Robert Wood Johnson Medical School, University of Medicine and Dentistry of New Jersey, Piscataway, New Jersey 08854, U.S.A.

Masando Hayashi, Department of Pathology, Robert Wood Johnson Medical School, University of Medicine and Dentistry of New Jersey, Piscataway, New Jersey 08854, U.S.A.

Helen H. Hess, Office of the Scientific Director, National Eye Institute, National Institutes of Health, Bethesda, Maryland 20892, U.S.A.

Leslie Hyman, Division of Epidemiology, School of Medicine, State University of New York at Stony Brook, Stony Brook, New York 11794-0836, U.S.A.

Jeffrey Jacobs, Departments of Surgical Research and Pathology, The Children's Hospital and Harvard Medical School, Boston, Massachusetts 02115, U.S.A.

Martin L. Katz, National Eye Institute, National Institutes of Health, Bethesda, Maryland 20892, U.S.A.

J. J. Knapka, Veterinary Resources Branch, Division of Research Services, National Institutes of Health, Bethesda, Maryland 20892, U.S.A.

Maureen B. Maude, Cullen Eye Institute, Baylor College of Medicine, Houston, Texas 77030, U.S.A.

Harold A. Mueller, Medical College of Virginia, Virginia Commonwealth University, Richmond, Virginia 23298, U.S.A.

Muna I. Naash, Cullen Eye Institute, Baylor College of Medicine, Houston, Texas 77030, U.S.A.

T. L. O'Keefe, Department of Physiology and Pharmacology, School of Veterinary Medicine, Purdue University, Lafayette, Indiana 47907, U.S.A.

Alicia Orlidge, Departments of Surgical Research and Pathology, The Children's Hospital and Harvard Medical School, Boston, Massachusetts 02115, U.S.A.

John S. Penn, Cullen Eye Institute, Baylor College of Medicine, Houston, Texas 77030, U.S.A.

Laurence M. Rapp, Cullen Eye Institute, Baylor College of Medicine, Houston, Texas 77030, U.S.A.

W. Gerald Robison, Jr., National Eye Institute, National Institutes of Health, Bethesda, Maryland 20892, U.S.A.

John C. Saari, Department of Ophthalmology, University of Washington School of Medicine, Seattle, Washington 98195, U.S.A.

Robert L. Trelstad, Department of Pathology, Robert Wood Johnson Medical School, University of Medicine and Dentistry of New Jersey, Piscataway, New Jersey 08854, U.S.A.

Rex D. Wiegand, Cullen Eye Institute, Baylor College of Medicine, Houston, Texas 77030, U.S.A.

J. S. Zigler, Jr., Laboratory of Mechanisms of Ocular Disease, National Eye Institute, National Institutes of Health, Bethesda, Maryland 20892, U.S.A.

Fibrils, Fibonacci and Fractals: Searching for Rules and Rulers of Morphogenesis in the Orthogonal Stroma of the Chick Cornea

Robert L. Trelstad, Masando Hayashi, Kimiko Hayashi, and David E. Birk

The chick cornea is a relatively simple structure. It forms the outer, exposed sector of the eye; it is avascular; it is transparent; it is comprised predominantly of one family of proteins, the collagens. The rope-like collagen molecules are woven into longer rope-like fibrils, collected into bundles, distributed in sheets disposed in orthogonal directions all which describe a spiral-like organization resembling a cholesteric liquid crystal (Coulombre, 1965; Trelstad and Coulombre, 1971; Trelstad, 1982b). The individual collagen fibrils in the chick are constructed from at least two and probably more different molecular species of collagen and are associated with several types of proteoglycans. These heteropolymeric fibrils are regularly spaced within the bundle/layer and are remarkably regular in diameter (Trelstad and Coulombre, 1971; Linsenmayer et al., 1984; Birk and Trelstad, 1984). The transparency of the cornea is dependent on the spatial order of the stroma.

The morphogenesis of the chick cornea has been described in expanding detail over the past two decades and represents one of the best studied tissues in the embryo (Hay and Revel, 1969; Hay et al. 1979; Hay, 1985). A series of diagrams depicting the macroscopic and microscopic features of corneal morphogenesis is presented in this chapter.

The reader should review these diagrams and consider the series of questions posed in the following paragraphs. An attempt will be made to integrate current information with unresolved questions for each of the diagrams. At the conclusion of this review, consideration will be given to appropriate strategies for explication of current knowledge as well as theoretical strategies for future studies.

CORNEAL MORPHOGENESIS

The budding of the lens vesicle from the head ectoderm and its closure and separation from the ectoderm mark the beginning of the corneal epithelium and the onset of the production by it of the primary corneal stroma, a cell free matrix which accumulates beneath the epithelium (Figures 1-4). Studies using isolated corneal epithelial cells and subsequent immunocytochemical studies all have established that the corneal epithelium is producing collagen types I and II (Linsenmayer et al., 1977; von der Mark et al., 1977). The epithelium also is producing fibronectin (Kurkinen et al. 1979) and the proteoglycans, chondroitin sulfate and heparan sulfate (Figure 5) (Meier and Hay, 1973; Trelstad et al., 1974). In future studies of the early commitment of epithelial cells of the head ectoderm to the 'corneal' phenotype, definitions will be restated in terms of the transcription and translation of the genes for collagen types I and II. Using in situ hybridization, it is now possible to detect small amounts of collagen type specific mRNA in the corneal epithelial cells at the earliest times of their appearance (Hayashi et al, 1986). In fact, preliminary data suggest that the 'corneal epithelial phenotype', as defined by the appearance of type II collagen mRNA, occurs in the head ectoderm even before separation of the lens vesicle (Hayashi et al., unpublished).

As the corneal epithelium continues to deposit the primary corneal stroma at stage 23 (Figure 3), the endothelium forms along its posterior surface (Hay and Revel, 1969). The endothelial cells will contribute to the subsequent development of the corneal stroma in a number of ways. It has been identified as the source of the hyaluronic acid which is abundant in the early development of the primary stroma (Toole and Trelstad, 1971). It forms a barrier to diffusion across the posterior surface of the cornea, ultimately leading to the stromal deturgescence necessary for transparancy (Coulombre, 1969). And recent studies (Hayashi, M. et al., unpublished) indicate that the endothelium also is contributing collagens to the primary stroma.

The role of the endothelium in the regulation of the ingrowth of the neural crest derived mesenchyme at the periphery remains unclear. The close temporal correlation between the production of hyaluronate, the formation of a complete endothelial layer, the swelling of the stroma and the infiltration by mesenchyme have

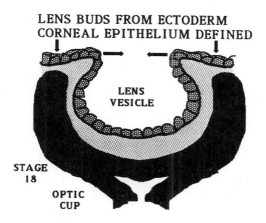

LENS BUDS FROM ECTODERM
CORNEAL EPITHELIUM DEFINED

LENS
VESICLE

STAGE
18

OPTIC
CUP

Figure 1. At stage 18 (Hamilton and Hamburger) the lens vesicle buds from the head ectoderm under the influence of the optic cup. The cells which will become the corneal epithelial cells are those head ectodermal cells which overlie the lens/optic cup.

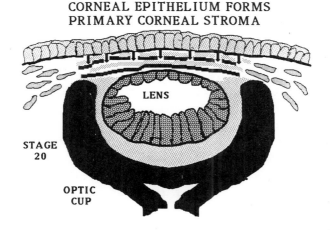

CORNEAL EPITHELIUM FORMS
PRIMARY CORNEAL STROMA

LENS

STAGE
20

OPTIC
CUP

Figure 2. At stage 20, the corneal epithelium is two cell layers thick and the basal layer is actively depositing the primary corneal stroma, comprised of collagen types I and II and chondroitin sulfate proteoglycan. The neural crest derived head mesenchyme do not participate in primary corneal stromal formation nor do they invade it until about stage 28.

4

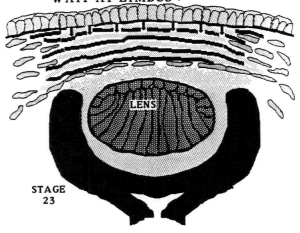

PRIMARY STROMA CONTINUES
TO BE DEPOSITED
ENDOTHELIUM FORMS
NEURAL CREST MESENCHYME
WAIT AT LIMBUS

LENS

STAGE
23

Figure 3. At stage 23, the primary stroma continues to accumulate beneath the outer epithelium. The endothelium forms from cells which migrate across the posterior face of the primary stroma. The crest derived mesenchyme continue to wait at the limbus.

supported the idea that these events are causally related (Toole and Trelstad, 1971; Toole et al., 1984). However, we do not have a detailed explanation of this 'transformation' of the head neural crest as it invades the primary corneal stroma.

While the extracellular matrix, in the cornea and elsewhere, is the scaffold on which the structure of the organism builds, it also serves as a soluble and solid phase receptor and agonist (Trelstad, 1985). The 'signalling' ability of the matrix might help to partly explain the 'transformation' of the head crest cells from non-infiltrative to infiltrative. Small peptides and oligosaccharides embedded within and/or released from the matrix can significantly affect cell synthesis, replication and migration (Malone et al., 1982; Castellot et al., 1984; Terranova et al. 1985; Majack and Bornstein, 1985; Goodman and Newgreen, 1985; Reddi, 1985; Ruoslahti and Pierschbacher,1986). Solid phase reactions between the cell and adjacent matrix are probably equally as regulatory. Thus, while the explanation given above of the invasion of the primary corneal

PRIMARY STROMA SWELLS
CREST CELLS INVADE & BEGIN
TO DEPOSIT SECONDARY STROMA
PRIMARY STROMA CONTINUES
TO BE DEPOSITED

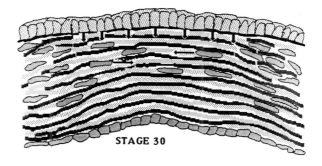

STAGE 30

Figure 4. At stage 30, the invasion of the primary stroma has already begun. The invading cells use the primary stroma as a scaffold and presumably as a template. During this period, the corneal epithelium continues to deposit more primary stroma.

stroma by crest cells uses a 'physical metaphor', in that the swelling of the stromal space provides a space within which the cells can migrate, we need to explore the 'signalling metaphor' in which the matrix components, indicated in Figure 5, play a direct role through changes in size, composition and/or conformation. As will be discussed later, it is possible that both metaphors are relevant and that there are redundant mechanisms responsible for this important phase of corneal development.

Little attention has been directed at the factors which regulate the number of cells which occupy the cornea. In addition to invasion from the limbus, there is also active replication of crest cells from day 5 through day 13 of development. The neural retina is a rich source of growth factors (Moczar et al., 1981) and it would seem reasonable to conclude that growth factors from this tissue and others might affect some of the events in the early cornea.

The scaffold which the head crest cells invade is organized in a manner best described as a spiralling orthogonal set of collagen layers closely resembling a cholesteric liquid crystal (Trelstad, 1982). This complex geometry in the primary stroma is essentially identical to the organization found in the adult stroma. The genesis

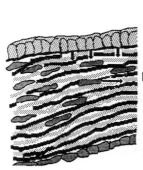

CORNEAL EPITHELIUM:
TYPE I COLLAGEN
TYPE II COLLAGEN
CHONDROITIN SO_4
HEPARAN SO_4
FIBRONECTIN

FIBROBLASTS:
TYPE I COLLAGEN
TYPE V COLLAGEN
KERATAN SO_4
DERMATAN SO_4

ENDOTHELIUM:
HYALURONATE

Figure 5. The matrix products produced by the epithelium, fibroblasts and endothelium for which there is substantial evidence are indicated. The products produced by the corneal epithelium become incorporated into the basement membrane of the epithelium as well as enter into the primary stroma. The way in which the products are 'sorted' and assembled separately is not clear.

of the cholesteric liquid crystal like architecture of the adult stroma is thus possibly dependent on the primary stroma in that the crest cells invade the primary stroma and replicate its pattern. As indicated in Figure 6, the architecture of the stroma is bilaterally asymmetrical whereas other structures immediately adjacent such as the scleral ossicles are bilaterally symmetrical.

What accounts for this handedness? A number of years ago it was suggested that corneal morphogenesis represents a macroscopic self assembly system (Trelstad and Coulombre, 1971). Because of the well known properties of isolated collagens to self assembly into ordered macroaggregates, (Trelstad, 1982a; Gross and Bruns, 1984) it seemed reasonable to conclude in 1971 that the extended play of physicochemical forces at successive stages of size and complexity could account for this unusual architecture.

We know that the individual collagen molecule has chirality or handedness (Piez, 1984). The triple helix, formed from the three collagen alpha chains within the cisterns of the endoplasmic reticulum, is right handed. Accordingly, fibrils must have a handedness and so must bundles, layers and tissues. While this simple analysis has some attraction, it fails to explain the fact that in most tissues there is bilateral symmetry rather than

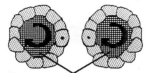

**THE SUBMAMMALIAN CORNEAL
STROMA RESEMBLES A CHOLESTERIC
LIQUID CRYSTAL. THE HANDEDNESS
OF THE STROMAL SHIFT IS
BILATERALLY ASYMMETRICAL WHILE
THE SCLERAL OSSICLES ARE
BILATERALLY SYMMETRICAL**

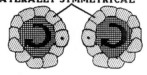

Figure 6. The images at the top and bottom of the diagram present a schematic en face view of of the two corneas of the chick. The thick arrows indiciate the common handedness of the shift in the orthogonal arrangement of the corneal stroma, proceeding clockwise from the outer to inner layers through 220 degrees. As indicated, the handedness of the stroma is the same in both eyes and bilaterally asymmetrical. The scleral ossicles are a series of bones which are present at the immediate periphery of the cornea. They are present in a complex overlapping pattern (Coulombre and Coulombre, 1973) which is bilaterally symmetrical.

asymmetry. It seems unlikely that the chirality of all macromolecules simply determine the chirality of all tissues. Complex structures such as ropes are woven from 'handed' sections, the 'handedness' alternating from one level to the next. Right handed 'threads' are wrapped about each other to form left handed 'ropes'; and left handed ropes wrap about each other to form right handed 'cables.' Hypothetically handedness could be controlled by a simple alternating switch which controls the step of the assembly sequence. This seems much too simple a rule to explain the many examples of handedness in biological systems, however.

THE COLLAGENS

The collagens are listed in Figure 7 and the top of the diagram illustrates several important features of these macromolecules. First, the collagen 'molecule' is, in the case of type I, a heteromonomer, that is, it is comprised of two different gene

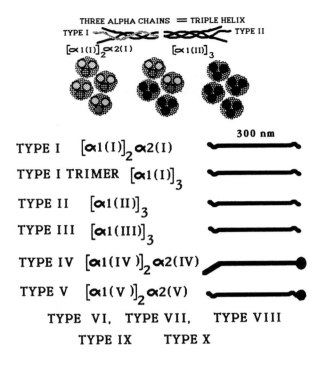

THREE ALPHA CHAINS = TRIPLE HELIX

TYPE I   TYPE II

$[\alpha 1(I)]_2 \alpha 2(I)$ $[\alpha 1(II)]_3$

300 nm

TYPE I $[\alpha 1(I)]_2 \alpha 2(I)$

TYPE I TRIMER $[\alpha 1(I)]_3$

TYPE II $[\alpha 1(II)]_3$

TYPE III $[\alpha 1(III)]_3$

TYPE IV $[\alpha 1(IV)]_2 \alpha 2(IV)$

TYPE V $[\alpha 1(V)]_2 \alpha 2(V)$

TYPE VI,   TYPE VII,   TYPE VIII

TYPE IX   TYPE X

Figure 7. The collagen molecule is comprised of three polypeptide chains or alpha chains wrapped into a right handed superhelix. This triple helical molecule may contain products of different genes such as is seen with type I collagen; or it may contain products of the same gene as seen with type II collagen. Cross sectional views of collagen molecules are illustrated below their longitudinal view. Each alpha chain is represented by a small circle; the triple helical molecule, by the enclosing larger circle. At the left, four collagen molecules of type I are present in a homopolymer of heteromonomers. In the middle, two type I and two type II collagen molecules are present within the same heteropolymeric 'fibril'. And at the right, a homopolymeric fibril comprised of homomonmeric type II collagen is illustrated. The list of collagen types I through V in the lower portions indicate the chain composition and nomenclature for these well characterized molecules. The approximate relative size for these collagens is indicated. The details of chain composition and size are not indicated for types VI through X.

products, alpha 1 type I and alpha 2 type I. In the case of type II, a homomonomer, all three collagen polypeptides are derived from the same gene for the alpha 1 type II chain. In humans, each of these genes is on a separate chromosome (Solomon et al., 1985; Ramirez et al., 1985). As can be seen from figure 7, collagen types I, IV and V are hetermonomers and types II, III and type I trimer are homomonomers. In the diagram at the top, below the longitudinal views of the triple helix, the triple helical collagens have been cut in cross section and the three alpha chains are indicated by the small filled circles and the molecule by the entire circular structure. Note that three different types of fibrils are shown: on the right are homopolymers of homomonomers, as would be seen with pure type II fibrils; in the middle are heterpolymers of homo and hetermonomers, for example a mix of type I and II in the same fibril; and on the left are homopolymers of heteromonomers, as would be seen with pure type I fibrils.

We know that the primary corneal stroma consists of heteropolymers of collagen types I and II and that the secondary stroma of types I and V (Hendrix et al., 1982; Linsenmayer et al., 1985; Birk et al., 1986). While we suspect that type IX is present in the primary stroma and VI in the secondary, their exact localization has yet to be established.

The conclusion that the architecture of the primary corneal stroma dictates that of the secondary corneal stroma derives from a simple comparison of their respective architectures. The similarity in three dimensional arrangement of these two structures in which both are organized like cholesteric liquid crystals, is a strong argument to support this conclusion. The morphogenesis of the secondary stroma would then be a problem of the 'translation' of the spatial information provided the invading neural crest cells into the products they deposit. That the neural crest cells use the primary stroma for anchorage during migration seems certain (Bard and Hay, 1975). That they obtain spatial information from this scaffold which will then dictate the architecture of the collagens they produce is the tentative explanation (Trelstad and Coulombre, 1971; Bard and Higginson, 1977).

However, if the primary stroma has sufficient 'information' within its macromolecules to effect the assembly of a three dimensional cholesteric crystal like structure, why can't that same potential information be present within the macromolecues which comprise the secondary stroma? Isn't it reasonable to argue that the similarities in architecture between the primary and secondary stromas merely reflect the inherent 'macroscopic self assembly' capacity of such matrixes? There are differences between the situation existent during the formation of the primary stroma and

that during the secondary: in the former there are no cells within
the forming stroma and the stroma is deposited along the relatively
planar surface of an epithelium; in the latter, cells occupy the
stroma as it forms and are not derived from one or two simple planar
faces. These differences are important, but were it to be shown that
a collagen type I/II mixture or a collagen type I/V mixture could
assemble into a three dimensional structure resembling the in vivo
stroma, does that mean that the primary stroma does not drive the
architecture of the secondary? Would this interesting in vitro
experiment, which has yet to be done, indicate that the formation of
the two structures is independent? Or might it not be possible,
despite the independent capacities of the two stromas, that the role
of the primary stroma is, nonetheless, to serve as a template for
that secondary stroma and that the inherent capacity of the
secondary stroma to assume this morphology independently might be a
'backup mechanism' or redundancy? Moreover, might not the tendency
of corneal fibroblasts to produce an orthogonal array be another
potential redundancy (Bard and Hay, 1975; Burke and Foster, 1985)?
There are many examples of redundancies in biological systems and it
would not be unexpected to find them operating at molecular and
cellular levels during the morphogenesis of the chick cornea.

COLLAGEN FIBRIL ASSEMBLY

With or without redundancy, how does the corneal epithelium
produce a matrix which it discharges beneath its basal surface and
which then assembles into a complex architecture comprised of
collagen fibrils, fibril bundles and fibril layers? The aggregation
of type I collagen into fibrillar structures has been described in
considerable detail (Piez, 1984), but we do not know much about the
simultaneous assembly of heteropolymeric structures comprised of
types I and II. Nor do we understand how the specialized basal
surface of the epithelium, covered by a basal lamina, is involved in
the fibril assembly process. By comparison with recent studies of
fibril assembly by fibroblasts, some general principles can be
suggested (Birk and Trelstad, 1984; Birk and Trelstad, 1986). There
most likely is a cooperative and complex interplay among molecular,
cellular and other forces which control fibril formation in situ
(Table 1). That self assembly operates at the molecular levels to
influence the packing of the molecules seems certain. The self

TABLE I

SELF ASSEMBLY:
REQUIRES NO ENERGY INPUT
MASS ACTION DRIVEN
MULTI-STEP
ENVIRONMENT DEPENDENT
POLYMORPHISMS
HOMO &/OR HETEROPOLYMERIC

CELL DIRECTED ASSEMBLY:
ENERGY DEPENDENT
COMPARTMENT DEPENDENT
TOPOGRAPHY DEPENDENT

ASSEMBLY REARRANGEMENTS:
NON-ENZYMATIC MODIFICATION
ENZYMATIC MODIFICATION
COMPONENT ADDITION
AGGREGATE FUSION
AGGREGATE REPOSITIONING

assembly process requires little or no energy input; is mass action driven; is usually a multi-step process; may lead to polymorphisms; and can involve molecules of the same type or of different types. These forces of 'self assembly' act in concert with forces of 'cell directed assembly' to create the first stage of matrix architecture. Cell directed factors are energy dependent and regulated by the compartments which the cell forms from intracellular membranous structures and extracellularly by the complex topography of the plasma membrane. Partitioning of secretory products prior to discharge in discrete intermediate assembly packages by the endoplasmic reticulum and Golgi apparatus are an important first stage in cell directed assembly; subsequent partitioning of the pericellular extracellular space into assembly sites where products are discharged is another important stage. Ultimately the assembly process is removed from the immediate extracellular compartments which the cells create and the matrix components are relatively free to undergo post-depositional rearrangements which include non-enzymatic and enzymatic modifications, additions of components either by aggregate fusion or the addition of diffusible precurors, and even aggregate repositioning by 'traction' mediated processes (Harris et al., 1981; Murray and Oster, 1984).

If we examine the corneal epithelial cells during the period when they deposit the primary corneal stroma we find that their shape and intracellular organization undergo remarkable changes. During the period from day 3 to 6, the perimeter of a basal corneal epithelial cell, viewed along the optic axis, undergoes an elongation from a 'length/width' ratio of 3 to one greater than 10 (Trelstad and Coulombre, 1971) (Figure 8). Abruptly on the 6th day, when the primary stroma is swelling and the neural crest derived cells invading, this elongation of the epithelial cells disappears and the cells remain isodiametric for the remainder of the animal's life. At this same time, the Golgi apparatus and centrioles within these same basal epithelial cells undergo a remarkable shift in intracellular position from the apical pole of the cell to the basal pole (Trelstad, 1970) (Figure 9). The significance of this shift would appear to be related to the secretory activity of these cells and the fact that the site of discharge of the matrix is across the basal surface of the cell. The reorientation of the cells' apical/basal axis reverts transiently on day 14 following which the Golgi once again moves to the basal cell pole; during this second shift, however, the centrioles remain in the 'apex' of the cell.

Are these architectural shifts of the cell important? There is no direct experimental evidence to bear on this argument. However, there is good evidence that the cytoskeleton is linked to the biosynthetic as well as morphogenetic functions of the corneal epithelium (Sugrue and Hay, 1981; Hay, 1985). In Figure 10 we have depicted the basal portion of a corneal epithelial cell in which we indicate the presence of secretory organelles, microtubles and actin filaments; possible linkages among them are indicated by the question marks. The cytoskeleton plays an important role in the orientation of the matrix components even before they are discharged from the cell. The absolute orientation of the primary stroma is the same from one animal to the next. The posterior layers are always parallel and at right angles to the corneal diameter which passes through the ventrally located choroid fissure. Given this fixed anatomic relationship of the geometry of the primary stroma and the choroid fissure axis of the eye there is little question that the organization of the primary stroma is closely coordinated with shaping reactions and morphogenesis in the rest of the organ and organism. Moreover, it suggests that the overall orientation of the matrix is provided by the corneal epithelium. Figure 10 poses another unresolved question which may or may not pertain to collagen assembly in the cornea. We do know that the type I and II molecules

DAY 3   6   7-ADULT

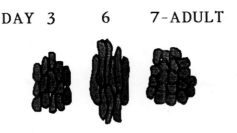

Figure 8. An en face view of the perimeter of the basal corneal epithelial cells at the days of development indicated. The choroid fissure axis in this diagram would be a horizontal line. The basal cells at day 3 are two to three times longer in one axis than the other. Over the next three days, the cells elongate or are elongated such that they are approximately ten times longer in one direction. On or about the late 6th day of development, the cells become isodiametric in shape and remain that way into the adult.

must traverse the epithelial basement membrane and that assembly occurs within or very close to the domain of the basement membrane. We do not know, however, whether type IV collagen is involved in this process nor whether it is present in the chick corneal basement membrane (Linsenmayer et al., 1984). Accordingly this diagram may not accurately reflect the exact corneal situation, but does illustrate, in a generic way, the problem of heteropolymeric collagen fibril assembly within an epithelial basement membrane.

STROMAL FIBROBLASTS AND MATRIX ASSEMBLY

Studies of tendon and corneal fibroblasts in the chick embryo have led us to propose that fibril assembly occurs in extracellular compartments which are formed by the convoluted topography of the cell's surface (Trelstad and Hayashi, 1979; Birk and Trelstad, 1984; Trelstad and Birk, 1984; Birk and Trelstad, 1986). In the epithelium, the surface is not convoluted and such extracellular compartments do not exist. Fibril assembly by the corneal epithelium is thus different from that for tendon and corneal fibroblasts.

When the fibroblasts invade the stroma they deposit the matrix which will be present in the adult. The assembly of this matrix occurs initially in close contact with the cell surface. Detailed ultrastructural studies of the stromal cells indicate that the initial steps in fibril assembly occur in compartments of the

GOLGI APPARATUS & CENTRIOLE
MOVEMENTS DURING FORMATION
PRIMARY CORNEAL STROMA

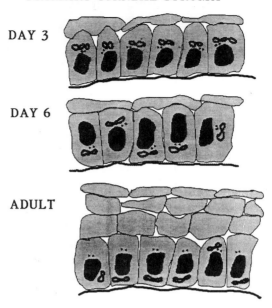

DAY 3

DAY 6

ADULT

Figure 9. A cross-sectional view through the corneal epithelium in a plane perpendicular to the corneal surface. The intracellular positions of the nucleus, Golgi apparatus and centrioles are indicated for days 3 and 6 of development and in the adult. At day three, the Golgi apparatus and centrioles are in their customary supranuclear position. By day six the Golgi apparatus and centrioles in the majority of cells are located in the basal cell pole. In the adult, the Golgi apparatus remains in the basal cell pole while the centrioles return to the apical cell pole.

extracellular space formed by the fusion of secretory vacuoles (Birk and Trelstad, 1984) (Figure 11). The subsequent assembly of fibrils occurs in this compartment, driven by forces of self assembly and by enzymatic processing. These fibril assembly compartments join to form a bundle assembly site within which the orientation and position of the fibrils in bundles is effected. The stromal fibroblast confronts a problem different from those of the tendon in that matrix must be deposited in two directions. How this is accomplished is not clear. Nor is it clear in what manner the pattern of the primary stroma is replicated. As we noted above,

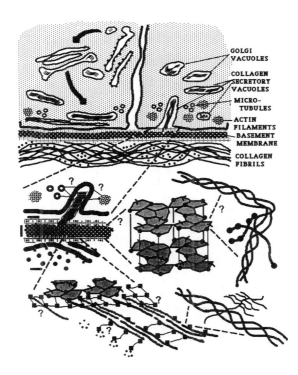

GOLGI
VACUOLES

COLLAGEN
SECRETORY
VACUOLES

MICRO-
TUBULES

ACTIN
FILAMENTS

BASEMENT
MEMBRANE

COLLAGEN
FIBRILS

Figure 10. The upper half illustrates the basal portion of the
corneal epithelial cell containing the intracellular compartments of
endoplasmic reticulum, Golgi apparatus and secretory vacuoles
responsible for the synthesis, glycosylation and transport of the
collagen to the cell surface. A basement membrane 60 nm thick
underlies the epithelium and the primary corneal stroma is
interwoven into the basement membrane. The dashed lines leading to
the lower left illustrate the basal cell membrane at higher
magnification and suggest that the secretory vacuole fused with the
cell membrane may (?) be linked to adjacent microtubules and/or
filament bundles. The dashed lines leading to the right from the
basement membrane indicate a possible packing arrangement for type
IV collagen derived from a model by Yurchenco et al. (1986). To the
right of that, at yet higher magnification, is a type IV collagen
molecule across which is draped a cross shaped laminin molecule
which is known to be in the chick corneal basal lamina (Meier and
Drake, 1984). The bottom of the diagram illustrates the molecular
interface between type I and II collagen molecules in an emerging
fibril and the outermost face of the basement membrane. At the very
bottom right a single collagen molecule with an interacting
proteoglycan is illustrated. The diagram intends to illustrate the
hierarchies of assembly from the molecular to the cellular in the
formation of the primary corneal stroma.

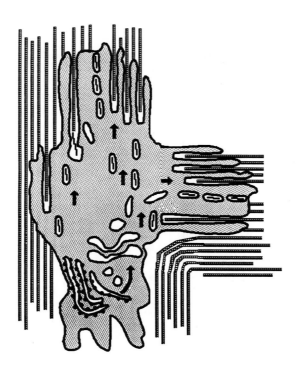

Figure 11. An idealized fibroblast in the secondary corneal stroma illustrating the intracellular and extracellular compartments formed by the fibroblast for the purposes of collagen fibril and fibril bundle assembly. Collagen is synthesized in the endoplasmic reticulum, moves to the Golgi apparatus and then through elongated secretory vacuoles to the cell surface where compound exocytosis leads to the formation of extracellular compartments surrounded by cytoplasm. The single, narrow compartments are the sites of fibril assembly; the lateral association of these compartments leads to bundle forming compartments. The simultaneous production of collagen in orthogonal directions is indicated.

perhaps the two systems are redundant. Assuming, however, that some form of replication occurs, how is the architecture of the primary stroma communicated to the fibroblasts and how is the replication effected? Might it be that there are a number of different assembly reactions, some mediated by the topography of the cell and accomplished within the extracellular compartments and others occuring by a diffusion mediated process? Might the process of template replication involve rapid deposition and remodeling such that the initial deposition can be error prone and the subsequent phases be driven by a stability conferred on those components which

**WHAT IS THE MECHANISM
OF TEMPLATE REPLICATION:
CELL TOPOGRAPHY ?
DIFFUSION/NUCLEATION?
REMODELING?
GNOMONIC?**

Figure 12. The fibroblasts which invade the primary corneal stroma encounter an orthogonal template. The matrix deposited by the fibroblasts mimics that of the primary corneal stroma. What is the mechanism(s) of template replication? Is the topography of the cell affected by the template such that the subsequent matrix is ordered? Does the fibroblast discharge collagen in a form which can diffuse and nucleate on the primary corneal stroma? Might the fibroblast produce collagen at random and only that which polymerizes on the template of the primary stroma be resistant to degradation or remodeling? Is the process gnomonic, that is, does the addition of new material increase the size of the template, but not change its shape?

by chance assembled properly? Is the process gnomonic (Figure 12)?

A gnomon is a geometric form which, when added to an another form increases the size of the initial form, but does not change its shape. A simple gnomon is illustrated in Figure 12. A more complex example of gnomonic growth is shown in Figure 13. In this figure and this particular example, the first step is not gnomonic, but the usefulness of this exception and example should become apparent. Two squares of dimension 1 are added together producing a rectangle with sides 2,1. A second square is drawn, whose side is equal to the sum of the two squares; a fourth square is then drawn whose side is equal to the sum of the third square and the second. The numbers in the center of each of the squares in the diagram is the length of the side of the square. In this sequence, the added gnomon in each case is a square whose size is the sum of the sides of the two preceding squares. At the bottom of the diagram, the successive stages have been positioned adjacent to each other to illustrate their general congruity and the gnomonic process of an increase in

18

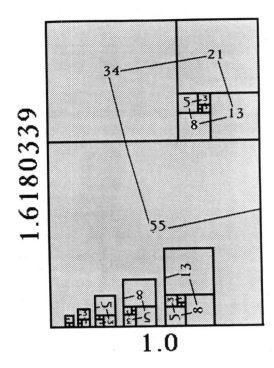

Figure 13. Near the upper right center to the right of the number 5 are two small squares. These two square abut another square to their right and the three squares form a rectangle. To this rectangle a fourth square is added whose side equals the sum of the second and third squares; to the '1,1,2,3' rectangle is added a square of side 5 and so on. The progression of adding squares whose side equals the sum of the two preceeding squares leads to a rectangle of dimension 1.0 X 1.6180339. This form is the 'golden rectangle' favored by ancient artisans for its perfect form. By joining the centers of each of the squares in the diagram an equiangular spiral can be formed. The diagam is an illustration of a Fibonacci series in which two numbers are added to obtain a third, the third and second to obtain a fourth and so on. The added square is a gnomon in this example in that it adds to a preexisting form without changing its shape. For ease of comparison, the growing rectangles are also illustrated across the bottom.

size with no change in shape. As the sequence progresses, a rectangle is generated whose dimensions are 1 to 1.6180339. The rectangle has been called the 'golden rectangle' because of its esthetic features and is found in a wide variety of architectural and art forms as well as in many natural structures (Huntley, 1970; Stevens, 1974).

FIBONACCI

The sequence of numbers illustrated in Figure 13 are elements of a Fibonacci series. This is a series in which any two numbers are added to create a third, the third added to the second to create the fourth and so on. The classical Fibonacci series of 1,1,2,3,5,8,13,21,34,55,89.... is well known to botanists because it precisely describes the number of left and right handed spiral rows in pinecones, sunflowers and other plant forms. The series is also found in animal forms in the spirals seen in nautilus shells and animal horns. An excellent discussion of the Fibonacci series and how it pertains to natural plant and animal forms is presented in Stevens (1974).

An unusual feature of the Fibonacci series is that after about 10 steps, regardless of the two starting numbers, the ratio of the last element to the next to last approaches 1.6180339 as a limit.

The number 1.6180339 and its negative reciprocal -0.6180339 also derive from the solution of the simple quadratic:

$$X^2 - X = 1$$

An unusual feature of the number .6180339 is that it provides a simple, progressive way to partition a space. The following program will successively subdivide a line of length 1 at position Z (where Z ranges from 0 to 1) such that after each step, the next subdivision will occur in the largest remaining space. As the value of X increments, the space between 0 and 1 will be partitioned by the value of Z in an 'equitable' manner, so that the segments formed are equal in size and the next division occurs in the largest remaining space. The value of Z 'points' to the site in space where that partitioning should occur and simultaneously subdivides that space in a controlled manner.

```
FOR X=1 TO 100
Y=X*(.6180339)
Z=Y-INT(Y)
'PRINT or PSET Z'
NEXT X
```

Space partitioning is essential in morphogenesis, but the rules governing this process are obscure. The recent discoveries of the homeotic genes, which are related to segment development in Drosophila and other forms, has pointed to specific signals which operate to effect pattern (Gehring, 1985). While there is little doubt that the genesis of form is rooted in the genes of the host, the problem of how this information is played out during development is not clear at all. There must be a genetic component, but the way in which gene activity couples to structural patterns is still unclear. Does it require timely production of 'signalling molecules'; does it involve a 'clock' or 'timing mechanism'; does it involve 'on/off' switches? It is likely to involve all of these as well as the control of the inherent structure of the reactants.

The explanation of form will thus depend heavily on the activities of genes which are transient and for which no structural component is generated. The architecture of a wall of bricks is defined by the architect and the mason, not by the chemistry of the bricks and mortar. While the local organization of the brick wall is dependent on 'brickness', the overall structure is not. We have little understanding of what the 'architect' and 'mason' of morphogenesis are. We suspect that the cell is an integral part of the matrix assembly process and that while we will learn much about the rules for local organization from a detailed examination of the primary structure of the reactants, we won't learn the entire story.

The genealogy of the male bee does not immediately come to mind when considering corneal morphogenesis, but such a genealogy is shown in Figure 14. Male bees are derived from unfertilized eggs while female bees are derived from fertilized eggs. The result of this simple difference is that the parental lineage of a male bee consists of a Fibonacci-like series. In Figure 15, a simple statement of several different rules for branching are defined and a structure derived from the application of this rule is shown.

Is there a place for the Fibonacci series in the analysis of fibril assembly or corneal morphogenesis? At present there is no reason to consider any meaningful linkage as there is, for example, in the morphogenesis of plants (Stevens, 1974). However, our capacity to recognize patterns is determined, in part, by our awareness of patterns and pattern variants as well as our ability to describe the complex geometries seen in nature. Simple Euclidean descriptions are inadequate. Students of morphogenesis must become more literate in the language of 'space'. Three recent volumes to which the reader is referred for more detail are those by Meinhardt (1982), Subtelny and Green (1982) and Malacinski and Bryant (1984).

There is one observation which we have made which links the prior paragraphs with the cornea in a very tenuous way. We comment

GENEALOGY OF THE MALE BEE

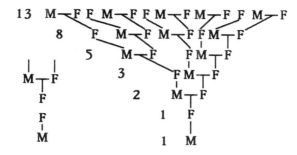

Figure 14. The genealogy of the male bee. Males derive from unfertilized eggs and thus have only one parent. Females derive from fertilized eggs and thus have two parents. The numbers of ancestors of a male bee describes a typical Fibonacci series.

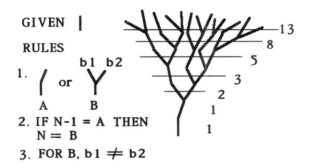

Figure 15. A rule set for the generation of a branching structure. Given a straight segment, such a segment can either extend one length or bifurcate. The action at each step must not be the same as the prior step. With a bifurcation, whatever happens to one limb (extend or bifurcate) cannot happen with the other limb. The consequence of these two rules determines a branching pattern in which the number of limbs at each level is an element of the classic Fibonacci series.

on it only to suggest how familiarity with space and pattern might alert us to linkages which might be explored. The angular shift of the cholesteric liquid crystal-like corneal stroma in the chick is consistently about 220 degrees. Is there anything unique about 220 degrees? Is it just coincidence that the product of 360 and .6180339 is 222? Is this trivial? Our increased awareness of the possibility that rules of space operate during morphogenesis as well as physicochemical rules of macromolecules, allow us to consider such observations with curiosity and caution.

FRACTALS

The description of complex biological forms using traditional Euclidean geometry is difficult because biological forms do not have regular shapes (Stevens, 1974; Loeb, 1976; Rucker, 1977). Fractal geometry provides, perhaps, a better way in which to describe biological forms and considerable recent attention has been directed toward this new area. The term fractal was coined by Mandelbrot and derives from the Latin fractus meaning 'irregular or fragmented' (Mandelbrot, 1983).

A simple way to compare fractal geometry and traditional Euclidean geometry is to consider the relationship between mass (M) and length (L).

The relationship:
$$M = KL^d$$
where K is a constant can describe a number of simple Euclidean forms such as a line (d=1), an area (d=2) and a volume (d=3).

The relationship
$$M = KL^D$$
where K is a constant and in which D is not an integer can describe forms which are not simple in their shape and which resemble that shown in Figure 16. In the cases where d=D, the system is Euclidean.

A second definition of fractals is that they describe forms which are 'self similar.' This can be seen in Figure 16 in which the shape of the smallest object is similar to the shape of each of the successively larger elements. In this example, the T shaped subunit is present at successively higher levels. The overall pattern resembles that seen in branching situations such as the lung.

Are fractals 'better' than Euclidean forms to describe biological structures? The world in which we live is not Euclidean. Are we to expect that the geometry of molecular and multimolecular aggregates does not conform to the 'real' geometry of the universe? Isn't it a strategic error to consider all forms of biological architecture in simple Euclidean terms?

The self similar definition of fractals and the architecture of the tendon provides an interesting 'verbal' fractal. The collagen molecule is a rope-like molecule comprised of three polypeptide

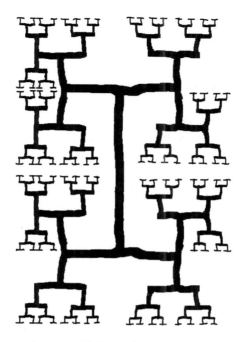

Figure 16. A fractal resembling a branching structure such as the lung. The figure is comprised of 'self similar' units with a T shape. Beginning with the major 'trunk' and two 'branches', the figure can be generated by adding successive branches with the new 'trunk' perpendicular to the prior branch (after Mandelbrot, 1983).

chains. The collagen fibril is a rope-like structure comprised of collagen molecules. The collagen fibril bundle is a rope-like structure comprised of collagen fibrils. The tendon is a rope-like structure comprised of collagen fibril bundles.

While this may simply be word play, it poses the question: are the ways in which we describe and attempt to understand the structure of the chick cornea sufficient? It is our opinion that they are not. The description of the submammalian corneal stroma as 'plywood like' is wholly inadequate. To liken it to a cholesteric liquid crystal changes the analogy, but does little more to provide an accurate description. The rules which govern liquid crystals are quite explicit, but these rules cannot be invoked simply because the cornea resembles such structures. It is our intention in the next several years to attempt a detailed description of the corneal stroma in the chick by serial reconstruction of transmission electron micrographs and by description in terms of a new language, possibly fractal geometry.

Our attention in the preceding paragraphs has been both the data pertaining to the cornea and a series of hypothetical frameworks within which we might begin to evaluate or understand complex patterning. While there are no current 'theories' of morphogenesis which are suitable for ocular development we consider it essential to look for rules and rulers of the development of form as we struggle with uncovering the details of tissue structure. Hopefully by expanding our descriptive and analytical capacities such as illustrated by the discussions of the Fibonacci series and fractals we will come to better descriptors and understanding of the complex geometry of the avian eye and other tissues.

A second purpose in discussing fibrils, Fibonacci and fractals within the same text on corneal morphogenesis is to pose the problem of evidence and explanation in redundant, heterogeneous biological systems. We should not expect that simple cause/effect explanations will emerge for many biological phenomena. When we look for causal factors in corneal morphogenesis, we should be prepared to accept multiple causes for one effect as well as one cause for multiple effects. For example, the 'one effect' of corneal transparency is the summation of multiple causes including the spacing and composition of the corneal macromolecules; the absence of vasculature; the relative anhydrous state and the curvature of the eye. Further, the spacing of the macromolecules may be determined by molecular factors, cellular factors and biomechanical factors. These factors may operate in some kind of hierarchy in which one is more dominant than another at certain stages. To focus on just one; or to emphasize just one is a logical error. In order to properly study and expand our knowledge of redundant and heterogeneous systems, it will be essential to continue to evaluate and develop the logical pathways between laboratory observations and subsequent interpretations. We anticipate the need to have more effective ways of moving back and forth from theory to evidence and from evidence to theory. This subject is essentially a consideration of the logic and philosophy of science and a thorough treatment is beyond the scope of this article, but is considered in detail in a number of recent publications (Glymour, 1980; Gardner, 1985; Eldredge, 1985; Levins and Lewontin, 1985).

REFERENCES

Bard JBL, Hay ED. 1975. The behavior of fibroblasts from the developing avian cornea. Morphology and movement in situ and in vitro. J Cell Biol 67:400-418.

Bard JBL, Higginson K. 1977. Fibroblast-collagen interactions in the formation of the secondary stroma of the chick cornea. J Cell Biol 74:816-827.

Birk DE, Trelstad RL. 1984. Extracellular compartments in matrix morphogenesis: Collagen fibril, bundle and lamellar formation by corneal fibroblasts. J Cell Biol 99:2024-2033.

Birk DE, Fitch JM, Linsenmayer TF. 1986. Organization of collagen types I and V in the embryonic chick cornea. Invest. Ophthal. 27:34-41.

Birk DE, Trelstad RL. 1986. Extracellular compartments in tendon morphogenesis: Collagen fibril, bundle and macroaggregate formation. 103:231-240.

Burke JM, Foster SJ. 1985. Corneal stromal fibroblasts from adult rabbits retain the capacity to deposit an orthogonal matrix. Dev Biol 108:205-253.

Campbell S, Bard JB. 1985. The acellular stroma of the chick cornea inhibits melanogenesis of the neural-crest-derived cells that colonize it. J Embryol Exp Morphol. 86:143-54.

Castellot JJ Jr, Beeler DL, Rosenberg RD, Karnovsky MJ. 1984. Structural determinants of the capacity of heparin to inhibit the proliferation of vascular smooth muscle cells. J Cell Physiol 120:315-25.

Coulombre AJ. 1965. Problems in corneal morphogenesis. Advances in Morphogenesis 4:81-97.

Coulombre AJ. 1969. Regulation of ocular morphogenesis. Invest Ophthalmol. 8:25-31.

Coulombre AJ, Coulombre JL. 1973. The skeleton of the eye. II. Overlap of the scleral ossicles of the domestic fowl. Dev. Biol. 33:257-67.

Eldredge, N. 1985. Unfinished Synthesis. Biological Hierarchies and Modern Evolutionary Thought. Oxford University Press, New York.

Gardner, H. 1985. The Mind's New Science. A History of the Cognitive Revolution. Basic Books, Inc. New York.

Glymour, C. 1980. Theory and Evidence. Princeton University Press, Princeton, New Jersey.

Gehring WJ. 1985. The homeo box: a key to the understanding of development? Cell. 40:3-5.

Goodman SL, Newgreen D. 1985. Do cells show an inverse locomotory response to fibronectin and laminin substrates?. EMBO J. 4:2769-71.

Gross J, Bruns R. Another look at fibrillogenesis. In: "The Role of Extracellular Matrix in Development", Trelstad RL, ed., A.R. Liss, New York, pp 479-512.

Harris AK, Stopak D, Wild P. 1981. Fibroblast traction as a mechanism for collagen morphogenesis. Nature. 290:249-51.

Hay ED, Revel J-P. 1969. Fine Structure of the developing avian cornea. In: Monographs on Developmental Biology. Wolsky A, Chen P, eds. Vol 1, New York, Karger, pp 1-44.

Hay ED, Linsenmayer TF, Trelstad RL, von der Mark K. 1979. Origin and distribution of collagens in the developing avian cornea. Current Topics Eye Res, pp 1-31.

Hay ED. 1985. Matrix-cytoskeletal interactions in the developing eye. J Cell Biochem. 27:143-56.

Hayashi M, Ninomiya Y, Parsons J, Hayashi K, Olsen BR, Trelstad. 1986. Differential localization of mRNAs of collagen types I and II in chick fibroblasts, chondrocytes, and corneal cells by in situ hybridization using cDNA probes. J Cell Biol. 102:2302-2309.

Hendrix MJ, Hay ED, von der Mark K, Linsenmayer TF. 1982. Immunohistochemical localization of collagen types I and II in the developing chick cornea and tibia by electron microscopy. Invest Ophthalmol Vis Sci. 22:359-75.

Huntley, HE. 1970. The Divine Proportion. Dover Publications, Inc., New York.

Kurkinen M, Alitalo K, Vaheri A, Stenman S, Saxen L. 1979. Fibronectin in the development of the embryonic chick eye. Dev Biol 69:589-600.

Levins R and Lewontin R. 1985. The Dialectical Biologist. Harvard University Press, Cambridge.

Linsenmayer TF, Smith GN Jr, Hay ED. 1977. Synthesis of two collagen types by embryonic chick corneal epithelium in vitro. Proc Natl Acad Sci USA. 74:39-43.

Linsenmayer TF, Fitch JM, Mayne R. 1984. Extracellular matrices in the developing avian eye: Type V collagen in corneal and noncorneal tissues. Invest Ophthalmol Vis Sci 25:41-47.

Linsenmayer TF, Fitch JM, Gross J, Mayne R. 1985. Are collagen fibrils in the developing avian cornea composed of two different collagen types. Ann NY Acad Sci 460:232-257.

Loeb AL. 1976. Space Structures. Their harmony and Counterpoint. Addison-Wesley Publishing Co, Reading, Mass.

Majack RA, Bornstein P. 1985. Heparin regulates the collagen phenotype of vascular smooth muscle cells: induced synthesis of an Mr 60,000 collagen. J Cell Biol 100:613-18.

Malacinski GM, Bryant SV. 1984. Pattern Formation. A primer in Developmental Biology. Macmillan Publishing Co, New York.

Malone JD, Teitelbaum SL, Griffin GL, Senior RM, Kahn AJ. 1982. Recruitment of osteoclast precursors by purified bone matrix constituents. J Cell Biol. 92:227-30.

Mandelbrot, BB. 1983. The Fractal Geometry of Nature. W.H. Freeman and Co., New York.

Meier S, Hay ED. 1973. Synthesis of sulfated glycosaminoglycans by embryonic corneal epithelium. Dev Biol. 35:318-31.

Meier S, Drake C. 1984. SEM localization of laminin on the basement membrane of the chick corneal epithelium with immunolatex microspheres. Dev Biol. 106:83-8.

Meinhardt H. 1982. Models of Biological Pattern Formation. Academic Press, New York.

Moczar E, Laurent M, Courtois Y. 1981. Effects of retinal growth factor and of the increase of the number of subcultures on sulfated glycosaminoglycans of bovine lens epithelial cells. Biochim Biophys Acta. 675:132-9.

Murray JD, Oster GF. 1984. Cell traction models for generating pattern and form in morphogenesis. J Math Biol. 19:265-79.

Piez, KA. 1984. Molecular and aggregate structures of the collagens. In: Extracellular Matrix Biochemistry. Piez KA, Reddi AH, eds. Elsevier, New York, pp 1-39.

Ramirez F, Bernard M, Chu ML, Dickson L, Sangiorgi F, Weil D, De Wet W, Junien C, Sobel M. 1985. Isolation and characterization of the human fibrillar collagen genes. Ann NY Acad Sci. 460:117-29.

Reddi AH. 1985. Implant-stimulated interface reactions during collagenous bone matrix-induced bone formation. J Biomed Mater Res. 19:233-9.

Rucker, RVB. 1977. Geometry, Relativity and the Fourth Dimension. Dover Publications, Inc., New York.

Ruoslahti E, Pierschbacher MD. 1986. Arg-Gly-Asp: a versatile cell recognition signal. Cell. 44:517-8.

Solomon E, Hiorns LR, Spurr N, Kurkinen M, Barlow D, Hogan BL, Dalgleish R. 1985. Chromosomal assignments of the genes coding for human types II, III, and IV collagen: a dispersed gene family. Proc Natl Acad Sci USA. 82:3330-3334.

Stevens, PS. 1974. Patterns in Nature. Little Brown and Co., Boston.

Subtelny S, Green PB. 1982. Developmental Order: Its origin and regulation. Alan R. Liss, Inc., New York.

Sugrue SP and Hay ED. 1981. Response of basal epithelial cell surface and cytoskeleton to solubilized extracellular matrix molecules. J Cell Biol 91:45-54.

Terranova VP, DiFlorio R, Lyall RM, Hic S, Friesel R, Maciag T. 1985. Human endothelial cells are chemotactic to endothelial cell growth factor and heparin. J Cell Biol. 101:2330-4.

Toole BP, Trelstad RL. 1971. Hyaluronate production and removal during corneal development in the chick. Dev. Biol. 26:28-35.

Toole BP, Goldberg RL, Chi-Rosso G, Underhill CB, Orkin RO. 1984. Hyaluronate-Cell Interactions. In: "The Role of Extracellular Matrix in Development", Trelstad RL, ed., A.R. Liss, New York, pp 43-66.

Trelstad RL. 1970. The Golgi apparatus in chick corneal epithelium: Changes in intracellular position during development. J Cell Biol 45:34-42.

Trelstad RL. 1982a. Multistep assembly of Type I collagen fibrils. Cell 28:197-198.

Trelstad RL. 1982b. The bilaterally asymmetrical architecture of the submammalian corneal stroma resembles a cholesteric liquid crystal. Dev Biol, 92:133-134.

Trelstad RL, Coulombre AJ. 1971. Morphogenesis of the collagenous stroma in the chick cornea. J Cell Biol. 50:840-58.

Trelstad RL, Hayashi K, Toole BP. 1974. Epithelial collagens and glycosaminoglycans in the embryonic cornea. J Cell Biol 62:815-830.

Trelstad RL, Hayashi K. Tendon fibrillogenesis: 1979. Intracellular collagen subassemblies and cell surface changes associated with fibril growth. Develop Biol 71:228-242.

Trelstad RL, Birk DE. 1984. Collagen fibril assembly at the surface of polarized cells. In: "The Role of Extracellular Matrix in Development", Trelstad RL, ed., A.R. Liss, New York, pp 513-543.

Trelstad RL. 1985. Glycosaminoglycans: mortar, matrix, mentor. Lab Invest 53:1-4.

von der Mark K, von der Mark H, Timpl R, Trelstad RL. 1977. Immunofluorescent localization of collagen types I, II and III in the embryonic chick eye. Develop Biol 59:75-85.

Yurchenco PD, Tsilibary EC, Charonis AS, Furthmayr H. 1986. Models for the self-assembly of basement membrane. J Histochem Cytochem. 34:93-102.

The Role of Matrix Components in the Control of Vascularization

Patricia A. D'Amore, Alicia Orlidge, and Jeffrey Jacobs

In the normal adult retina, the capillary endothelium seldom divides. Engerman and his co-workers (1967) have reported that the labelling index of the adult retinal vasculature is less than 0.01%; evidence that vascular growth is closely regulated under normal conditions. Yet, pathologic neovascularization is a complication of a number of ocular diseases including the retinopathy of prematurity and diabetic retinopathy.

Vascularization is undoubtedly influenced by a variety of factors, yet the specific mechanisms of growth control are poorly understood. Thus, the investigation of the events associated with capillary proliferation may provide valuable insight into potential sites of growth control and their role in microvascular pathologies.

Based on observations of newly formed blood vessels in the chick chorioallantoic membrane, Ausprunk and Folkman (1977) described specific processes associated with capillary formation. As depicted in Figure 1, one of the initial phases of capillary growth involves the dissolution of the basement membrane of the parent vessel, usually a venule, allowing the endothelial cell (EC) to break through the normally effective barrier. This process is thought to occur through the release of EC proteolytic enzymes, which act to break down the subjacent basement membrane. The EC then migrate through the disrupted basement membrane into the surrounding interstitial space. At the same time, the EC located behind the migrating front of cells undergo active proliferation. As the growing vessel approaches maturation, a lumen forms, pericytes arrive and blood flow commences.

The arrival of the pericyte or mural cell is believed to mark the maturation of the newly formed capillary (Crocker et al, 1970),

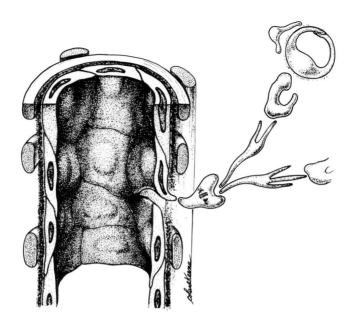

Figure 1: Schematic diagram of new capillary formation including the dissolution of the basement membrane of the parent vessel, the migration and proliferation of the EC, lumen formation, production of a new basement membrane and the arrival of the pericyte. Adapted from the studies of Yamagami (1970) and Ausprunk and Folkman (1977).

a stage that is also characterized by a shift in extracellular glycosaminoglycan (GAG) populations (Ausprunk, 1982). Both morphological analyses and clinical observations strongly indicate that the intimate association between pericytes and EC contributes to capillary integrity. The retinal microvasculature, which has one of the lowest turnover rates, also has the highest ratio of pericytes to EC (1:1), suggesting a role for the interaction of the two cells in growth control. Thus, it is interesting to note that the proliferation of retinal capillaries in diabetic retinopathy is temporally correlated with the "drop-out" of pericytes. Although the functional interaction between the capillary EC and pericyte is poorly understood, their physical association is well characterized and unique in that they share a basement membrane (Figure 2) and actually maintain points of plasma membrane contact. This is in contrast to the rest of the vasculature where the EC are clearly

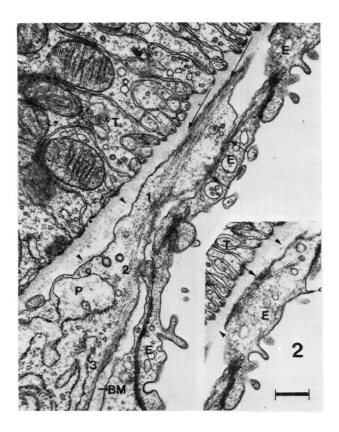

Figure 2: Transmission electron micrograph taken from the work of Courtoy and Boyles (1983) Journal of Ultrastructural Research. This illustrates the intimate association between capillary EC (E) and pericyte (P). A dense plaque (long arrows) of microfilaments is associated with the point of contact between the basement membrane of the pericyte and the tubule (T). These filaments are seen through three densities (1,2,3) in parallel with the endothelium. Plaque 2, in particular, is associated with a point of membrane contact between E and P. Short arrows illustrate points of close apposition, which are not associated with filaments. A well organized basement membrane (BM) is obvious in lower left. The insert is a continuation of the upper right corner of Figure 2 which illustrates EC filaments associated with dense plaques (arrows) at point of association of capillary and tubule basement membranes. (Bar=0.5 μ)

separated from the medial layer by a distinct basement membrane, and only contact the subjacent smooth muscle cells (SMC) under pathologic conditions.

Several factors have been implicated as potential regulators of neovascularization; these include the extracellular matrix, growth factors, cell-cell interactions and nutrient availability.

MATRIX AND VASCULARIZATION

The extracellular matrix, as well as specific matrix components, have been reported to influence various cell functions in vitro (Hay, 1981; Kleinman et al., 1981). The extracellular matrix has been shown to vary both quantitatively and qualitatively from tissue to tissue. Sage and Bornstein (1982) analyzed the extracellular matrix of EC cultured from different locations and found that the cells secreted a collagen type that reflected their origin. They concluded that these cells maintained differences in growth, morphology, migration and response to exogeneous factors through their interaction with the extracellular matrix. Interactions between cells and the matrix have been shown to modify growth (Gospodarowicz et al, 1978; Gospodarowicz et al., 1980; Kleinman et al, 1981), migration (Madri & Williams, 1983; Young & Herman, 1985) and differentiation (Hay, 1985; Kujawa & Tepperman, 1983; Lander et al, 1982). Using the amnion as a source of both native basement membrane and stromal matrix, Madri and Williams (1983) examined the effect of these matrices on the behavior of capillary EC. They found that EC plated onto basement membrane did not proliferate or migrate but "differentiated" into capillary-like tubes. Growing cells on interstitial matrix yielded cells that actively migrated and proliferated. Young and Herman (1985) examined the effect of different substrates on the response of aortic EC to injury in vitro and found that the cells' stress fiber number was inversely related to the rate of migration and that both variables were modulated by extracellular matrix components. Pratt and his co-workers (1984) reported changes in the distribution of an EC cytoskeletal protein (fodrin) in vitro as a function of matrix components.

Glycoproteins

Among the isolated matrix components that have been shown to influence cell behavior, much attention has been focused on the glycoproteins. Collagen represents a ubiquitous component of the extracellular matrix and is localized in the lamina densa of the basement membrane, a specific matrix located between cells and connective tissue stroma (Figure 3). Cells that are known to grow on a basement membrane in vivo preferentially attach to type IV collagen in vitro; whereas, cells isolated from interstitial spaces adhere to collagen types I and III in vitro (Murray et al., 1979; Madri & Williams, 1983). It is currently accepted that cells interact with collagenous substrates through other matrix-associated glycoproteins. In particular, fibronectin and laminin have been shown to mediate the attachment of several cell types to various surfaces including collagen and tissue culture plastic (Terranova et al, 1980; Grinnell et al., 1980). Both molecules are characterized by specific domains which mediate binding to cell surfaces as well as to other matrix components including collagen and GAGs (Yamada & Olden, 1978; Yamada et al., 1980, a & b).

Glycosaminoglycans

Recently, there has been a great deal of interest in the role of the GAGs in the modulation of various cell functions. GAGs are anionic polysaccharides composed of repeating units of hexosamine and uronic acid. They occur in vivo as proteoglycans or aggregates of polysaccharide chains on a core protein (with the exception of hyaluronic acid (HA), which exists in linear form). The GAGs have been shown to be associated with the cell surface (Lindahl & Hook, 1978) and have thus been implicated in cell-cell or cell-substrate interactions including attachment (Abatangelo et al., 1982; Barnhart et al., 1979; Stamatoglou and Keller 1983), motility (Majack & Clowes, 1984; Pratt et al., 1984; Forrester & Wilkinson, 1981), differentiation (Kujawa & Tepperman, 1983; Toole & Gross, 1971; Toole, 1981) and proliferation (Kawakami & Terayama, 1981; Matuoka & Mitsui, 1981, Castellot et al., 1981; Castellot et al., 1982; Clowes & Karnovsky, 1977).

Alterations in cell surface GAGs are associated with the modulation of cell growth in several systems. Neovascularization is correlated with significant shifts in the profile of basement membrane-associated GAGs. In the immature capillary, which is characterized by the absence of pericytes and extensive EC proliferation and migration, the dominant GAG is hyaluronic acid (HA). Upon maturation, the EC become quiescent and are in intimate

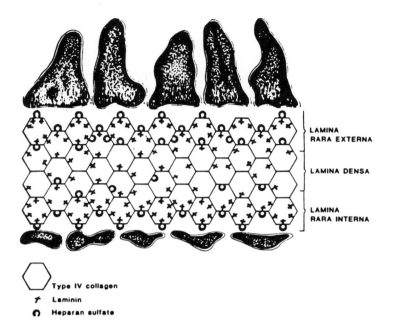

Figure 3: Schematic representation of the proposed organization of the glomerular basement membrane taken from the work of Martinez-Hernandez and Amenta (1983). Type IV collagen is distributed in all the layers of the basement membrane and forms the backbone for attachment of other components. Laminin and heparan sulfate proteoglycan predominate in the laminae rarae.

association with the newly-arrived pericytes. Concomitant with this, HA is degraded and is replaced by heparan sulfate (HS) (Ausprunk et al., 1982).

Glycosaminoglycans and Proliferation

Consistent with these observations in the vasculature are findings in several other cell systems. Kraemer and Tobey (1972) examined the changes in cell surface HS with respect to the cell cycle and found that it was selectively shed immediately prior to mitosis. Low density cultures of mammalian cells which are characterized by exponential cell growth are reported to synthesize higher levels of HA than confluent cultures (Tomida et al., 1974;

Underhill & Keller, 1976) which produce predominantly HS and chondroitin sulfate. The involvement of HS in growth control is indicated by in vitro studies correlating cell surface HS with the density-dependent inhibition of human fibroblast growth (Matuoka & Mitsui, 1981). In addition, using HS isolated from plasma membranes of quiescent confluent hepatocytes, Kawakami and Terayama (1981) induced contact inhibition in cultures of rapidly growing hepatoma cells.

Glycosaminoglycans and Attachment

Many studies have documented that substrate attachment is a prerequisite for appropriate growth and differentiation. Following the identification of GAGs as a major component of cell adhesion sites, Culp and colleagues (1979) proposed that the competitive binding of available proteoglycans to fibronectin mediates the appropriate interaction between a cell and its substrate. The GAGs, HA and heparin, are reported to influence cell-substrate interactions in several systems. In studies designed to elucidate the role of GAGs in cell attachment, Schubert and co-workers isolated from myoblast cultures a 16S particle composed of GAGs and glycoproteins that is involved in cell adhesion to a substrate (Schubert & LaCorbiere, 1980). Complexes of protein and GAGs known as "adherons" are released from chick neural retina cells in vitro and are reported to mediate the adhesion of these cells to a substrate (Cole et al., 1985). Recently, the GAG component of the adheron has been identified as a heparan sulfate proteoglycan, which clearly implicates this GAG in the mediation of attachment-dependent cell functions. In addition, the remaining polypeptide component of adherons is reported to have a distinct heparin-binding domain.

Variations in the level of extracellular HA, in particular, are associated with marked alterations in adhesion of several cell types. Increasing the level of exogenous HA was shown to facilitate the detachment of confluent monolayers of 3T3 cells (Abatangelo et al., 1982), suppress the directed and random movement of leukocytes (Forrester & Wilkinson, 1981), and reversibly inhibit myogenesis in primary cultures of 11-day chick muscle (Kujawa & Tepperman, 1983). Thus, the level of exogenous HA is clearly associated with the inhibition of cell-substrate interactions in vitro. Fisher and Solursh (1979) have suggested that similiar relationships exist in vivo with the enhancement of cell detachment functioning to facilitate cell migration of the HA-enriched matrices associated with morphogenesis (Toole, 1981), wound healing (Abatangelo et al., 1983), and possibly, vasculogenesis.

Glycosaminoglyans and Microvascular Cells In Vitro

In light of these observations, we were interested in studying the role of GAGs in the modulation of new blood vessel growth. To that end, we examined the effect of purified GAGs on the attachment and proliferation of vascular wall components--capillary EC, pericytes and aortic SMC; using retinal pigment epithelial cells and dermal fibroblasts as controls. Cell attachment was measured by comparing the number of cells that attached to GAG-coated subtrates with cell attachment to uncoated or bovine serum albumin (BSA)-coated tissue culture plastic at 2 hr intervals over an 8 hr period.

The effect of GAGs on cell proliferation was measured in response to the addition of purified GAGs to the culture media. For these experiments, cells were plated into 24 well tissue culture plates, allowed to attach overnight and then refed with the GAGs in fresh medium. After the time required for three population doublings (which varied with the cell type), the cells were enzymatically removed and counted electronically. Results are expressed as a percent of the number of control cells (incubated in media alone).

HA-coated substrates inhibited the attachment of both microvascular pericytes (Figure 4) and aortic SMC (Table 1) by up to 80% after 8 hr. Inhibition was observed by 2 hr and was consistent over the remaining period. The attachment of EC was not impaired by HA-coated substrates. On the contrary, HA enhanced the attachment of microvascular EC by 40% (Figure 4). The attachment of epithelial cells and fibroblasts was unaffected by HA-coated surfaces (Table 1). In addition, coating the tissue culture surface with heparin, chondroitin sulfate-C or chondroitin sulfate-B did not alter the attachment of any cell type examined (Figure 4 and Table 1).

The most significant effect of GAGs on cell proliferation was the dose-dependent inhibition of pericyte (Figure 5) and SMC (Table II) growth by heparin. Inhibition was evident at concentrations as low as 10 ng/ml and was maximal at 100 µg/ml. In contrast to the inhibition of proliferation observed with pericytes and SMC, heparin (100 µg/ml) enhanced the proliferation of capillary EC by a small

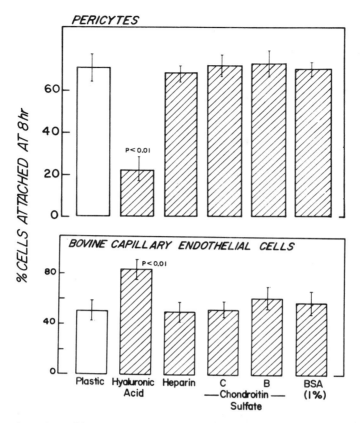

Figure 4: The effect of GAG-coated substrates on the attachment of capillary endothelium and pericytes. Cells were plated onto tissue culture wells that had been treated with 100 µg of GAG in 1 ml of PBS. Controls consisted of cells plated on untreated and BSA-coated tissue culture wells. The number of attached cells was determined after an 8 hr incubation. In this representative experiment, each bar corresponds to the average of quadruplicate cell counts.

but significant amount (Figure 5). No other GAG tested had any effect on the proliferation of any cell studied (Table II).

These data indicate that the GAGs, heparin and HA, have cell-specific effects on the attachment and proliferation of the cells of the microvasculature. Specifically, HA inhibits the attachment of pericytes and SMC, enhances capillary EC attachment, but does not alter the attachment of epithelial cells or fibroblasts. Heparin significantly inhibits the proliferation of pericytes and SMC, potentiates capillary EC growth, but does not affect the

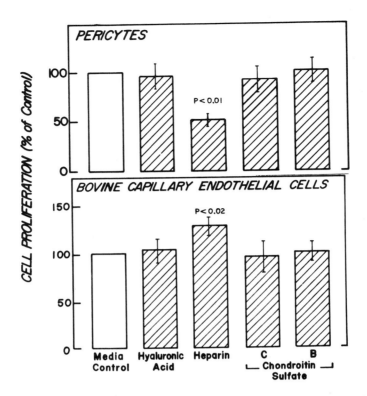

Figure 5: The effect of soluble GAGs on the proliferation of capillary endothelium and pericytes. Cells were plated onto tissue culture wells and allowed to attach overnight. Each GAG (1 ng/ml – 1 mg/ml PBS) was added to the cells in fresh media and incubated for the time required for three population doublings. Control cells received PBS with fresh media. Results are expressed as the percent change in the number of GAG-treated cells from the control cell number. In this representative experiment, each bar corresponds to the average of quadruplicate cell numbers.

proliferation of epithelial cells or fibroblasts.

GROWTH FACTORS IN VASCULARIZATION

Close to 30 years ago, George Wise (1956) postulated the involvement of growth factors in retinal neovascularization with the following statement:

EFFECT OF GLYCOSAMINOGLYCAN–COATED SUBSTRATES ON CELL ATTACHMENT

	capillary endothelial cells	pericytes	smooth muscle cells	fibroblasts	epithelial cells
heparin	0	0	0	0	0
hyaluronic acid	+	–	–	0	0
chondroitin sulfate C	0	0	0	0	0
chondroitin sulfate B	0	0	0	0	0
bovine serum albumin	0	0	0	0	0

Data obtained following an 8 hr incubation.

0 = no effect
– = inhibition of attachment
+ = enhancement of attachment

Table I.   The effect of GAG-coated substrates on cell attachment. The data depicted in this table were generated in the same manner as described in Figure 4.   In this case, the attachment of SMC, retinal pigment epithelial cells and dermal fibroblasts was also measured.

EFFECT OF EXOGENOUS GLYCOSAMINOGLYCANS ON CELL PROLIFERATION

	capillary endothelial cells	pericytes	smooth muscle cells	fibroblasts	epithelial cells
heparin	+	–	–	0	0
hyaluronic acid	0	0	0	0	0
chondroitin sulfate C	0	0	0	0	0
chondroitin sulfate B	0	0	0	0	0
bovine serum albumin	0	0	0	0	0

Data were obtained from the addition of 100µg/ml of each GAG.

0 = no effect
– = inhibition of attachment
+ = enhancement of attachment

Table II.   The effect of soluble GAGs on cell proliferation.   The data presented in this Table were generated in the same manner as described in Figure 5.   In this case, the proliferation of SMC, retinal pigment epithelial cells and dermal fibroblasts was also measured.

Pure retinal neovascularization is directly related to a tissue state of relative anoxia. Under such circumstances an unknown factor x develops in this tissue and stimulates new vessel formation, primarily from the capillaries and veins. In embryological development the gradient of this factor is orderly and preordained, and the development of the resulting retinal circulation is orderly and purposeful. When the definitive retinal vascular tree is completed, the production of factor x and its resulting new vessel formation ceases, and a state of equilibrium is reached. The capacity to form new blood vessels does not disappear from the capillaries and veins at this point: it simply lies dormant. In the adult retina, any condition bringing about a state of relative retinal anoxia first calls forth a compensatory retinal vessel dilatation with its resultant improved tissue oxygenation. Should this mechanism not satisfy the need, the dormant capacity of capillaries and veins to form new blood vessels is awakened and neovascularization results.

There is a great deal of indirect evidence that indicates a role for growth factors in the development of ocular neovascularization in pathologies such as retinopathy of prematurity or diabetic retinopathy. According to the hypothesis outlined above, retinal ischemia leads to the liberation of a stimulus that can elicit new blood vessels. Consistent with this concept is the fact that the current treatment for retinal neovascularization is laser destruction of non-perfused retina, the presumed source of the vasoproliferative factor. Additionally, the diffusion of an angiogenic factor from the retina has also been implicated in the development of iris neovascularization (Ashton, 1961).

For the past several years, we have been investigating the hypothesis that the retina contains a factor that can elicit new blood vessels. Because the available in vivo assay systems are tedious and not quantitatable, it was necessary to use in vitro systems to monitor the purification of the factor. These assay systems are based on the various events known to be involved in the formation of new blood vessels (Yamagami, 1970; Ausprunk and Folkman 1977), (see Introduction and Figure 1). We have used the proliferation of capillary EC as a screen for stimulatory activity during our purification process. We have also used the stimulation of DNA synthesis by BALB/c 3T3 cells to quantitate stimulatory activity. This sensitive assay system has been used extensively in the purification of other growth factors. We (D'Amore & Klagsbrun, 1984) and others (Shing et al., 1984) have found that growth factors that stimulate the proliferation of EC also stimulate DNA synthesis in 3T3 cells. Furthermore, since we know that the proliferation of capillary EC is not synonomous with the process of angiogenesis, we have also measured the angiogenic potential of the purified fractions using an in vivo assay system, the rabbit corneal pocket assay. Using these systems, we have purified from bovine retinas a polypeptide growth factor, retina-derived growth factor (RDGF), that is mitogenic for capillary EC in vitro and is angiogenic in vivo.

Identification and Characterization

In earlier studies, we demonstrated that isotonic buffer extracts of retinas stimulated the proliferation of aortic EC as well as the growth of new blood vessels on the chick chorioallantoic membrane (Glaser et al., 1980). Characterization of the stimulatory activity by physical, chemical and enzymatic treatment revealed that the factor was sensitive to heat, extremes of pH and protease treatment (D'Amore et al., 1981). These findings, together with the fact that the activity was retained after dialysis in tubing with a molecular weight cut-off of 14,000, indicated that the factor was a protein greater than 14,000 daltons. Ultrafiltration studies and gel exclusion chromatography of the crude retinal extract suggested a molecular weight of greater than 50,000 for the active material (D'Amore et al., 1981). Isoelectric focusing revealed that the active component was anionic with an isoelectric point of 4.7-5.2 (D'Amore & Klagsbrun, 1984).

Purification

Although some degree of purification was afforded by conventional biochemical techniques, efforts to purify the active component to homogeneity were hampered by the small amount of activity present in the retinas and by the large losses suffered during the purification process. A major breakthrough in the purification of RDGF occurred when we discovered that RDGF bound with high affinity to heparin. Shing and his co-workers (1984) had reported that a tumor-derived EC mitogen bound with high affinity to the glycosaminoglycan heparin, thus providing a powerful purification step. The tumor-derived mitogen was cationic, and therefore could be expected to bind to heparin. But, since we had already demonstrated that the factor was anionic, we thought it unlikely that the growth factor would bind to another anionic molecule such as heparin. Thus, we were surprised to find that all of the EC stimulatory activity in a crude retinal extract bound to heparin. Figure 6 illustrates the profile obtained from the application of a crude retinal extract to a column of immobilized heparin. More than 90% of the protein (dashed line) that was applied to the column did not bind to the heparin and was eluted in the void volume. Nearly all of the remainder of the protein eluted in the beginning of the salt gradient (before 0.6 M NaCl). However, all of the EC (open triangles) and 3T3 (closed circles) cell stimulatory activity eluted in a single peak with about 1.0 M NaCl.

The molecular weight of RDGF was approximated by applying heparin-affinity purified RDGF to a size exclusion HPLC column. Results of this step revealed that the EC and 3T3 cell stimulatory activity had a molecular weight in the range of 15-20,000. More recently, we have succeeded in purifying RDGF to homogeneity and have found that the active material is present in two forms with molecular weights of 18,000 and 16,500, respectively. The earlier estimate of a larger molecular size may reflect the existence of a carrier protein similar to that reported for other peptide growth factors such as epidermal growth factor (EGF).

Angiogenic Activity

The angiogenic activity of the highly purified RDGF was examined using the rabbit corneal pocket assay. This assay utilizes the avascular rabbit cornea as the vehicle in which to quantitate new blood vessel growth (Gimbrone et al., 1974). In this assay, a mid-corneal incision is made in the cornea of an anesthesized rabbit eye. The pocket is then created by inserting a spatula between the layers of the corneal stroma. To release the factor over an extended time, the RDGF was incorporated into pellets of ethylene vinyl acetate, a sustained release polymer (ELVAX) (Langer, 1981). A pellet containing RDGF or a control substance is then inserted into the pocket between 1 and 2 mm from the limbus. The response to the pellet is then monitored every other day for two weeks and the length and the density of the vessels are noted. The upper panel of Figure 7 shows a corneal pocket containing a pellet that contained

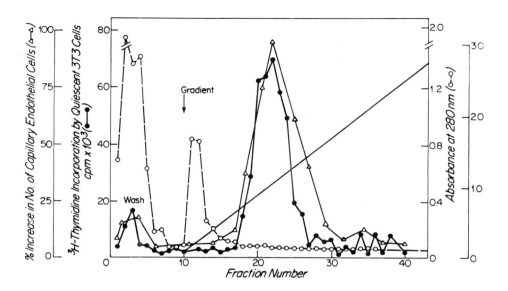

Figure 6: Heparin-Sepharose chromatography of RDGF. Crude retinal extract (59 ml) was dialyzed against 0.1 M NaCl, 0.01 M Tris-HCl, pH 7.4 and applied to a column of heparin-Sepharose. After a 50 ml rinse with the equilibration buffer, the proteins were eluted from the column with a gradient of 0.1-2.0 M NaCl. The fractions were dialyzed against deionized, distilled water and tested for their growth factor activity on 3T3 cells and capillary endothelial cells. All of the activity elutes from the column as a single peak with 1.0 M NaCl.

the control, rabbit serum albumin. Twenty days following implantation, no vessel growth is evident. The lower panel shows the neovascular response to a pellet that contained approximately 50 ng of highly purified RDGF. Histological observation of the cornea revealed the absence of inflammatory cells, indicating that the new vessels were the direct result of RDGF.

Retina-derived Growth Factor (RDGF)-Heparin Interaction

The binding of RDGF to heparin has been important for two reasons. First, the strong affinity of the factor for heparin has greatly facilitated its purification. The SDS polyacrylamide gel illustrated in Figure 8 graphically illustrates this point. The

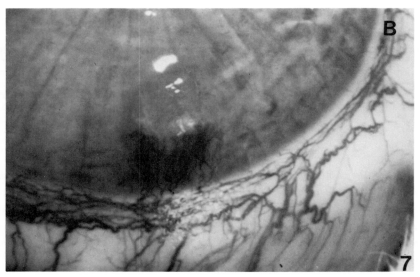

Figure 7:   Assay of the angiogenic activity of RDGF.   Panel (a)
shows the response in a rabbit corneal pocket to the control
carrier, rabbit serum albumin, 20 days after the pellet was
implanted.   No vessel growth is evident.   Panel (b) shows the strong
neovascular response to about 50 ng of highly purified RDGF 20 days
after the pellet was implanted.

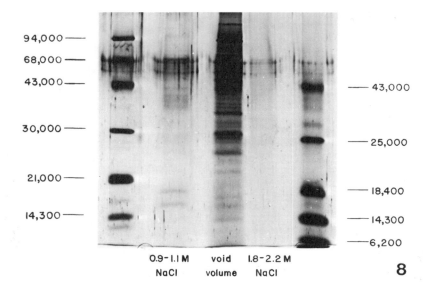

94,000 —
68,000 —
43,000 —
30,000 —
21,000 —
14,300 —

— 43,000
— 25,000
— 18,400
— 14,300
— 6,200

0.9-1.1 M   void   1.8-2.2 M
NaCl   volume   NaCl

8

Figure 8:   Silver-stained SDS polyacrylamide gel electrophoresis of
heparin-purified RDGF.   The lanes marked "void volume" contain the
proteins that did not bind to the heparin.   The lane labelled "0.9-
1.1 M NaCl" is the protein associated with the stimulatory activity
and contains only a few major bands, illustrating the power of
heparin-affinity chromatography as a purification scheme.

lane marked "void volume" contains the numerous proteins that did
not bind to the heparin.   In contrast, the lane labelled "0.9-1.1 M
NaCl" containing the fractions that were associated with the EC
stimulatory activity, reveals significantly fewer bands.   One
passage over a heparin-affinity column yields a 20,000-fold
purification of RDGF with 30-50% recovery and thus represents a
powerful purification step.

   The second level of significance of the RDGF-heparin affinity
lies in its potential physiological relevance. Three lines of
evidence suggest that the interaction between the growth factor and
heparin is a specific one.   Both heparin and chondroitin sulfate are
anionic GAGs with 1.6-3 and 0.1-3 sulfates per disaccharide,
respectively (Lindahl & Hook, 1978).   If the binding between RDGF
and heparin is based solely on ionic interactions then the factor
should also bind to chondroitin sulfate.   We have found, however,
that the EC stimulatory activity that binds with high affinity to
heparin does not bind at all to chondroitin sulfate (Figure 9),

46

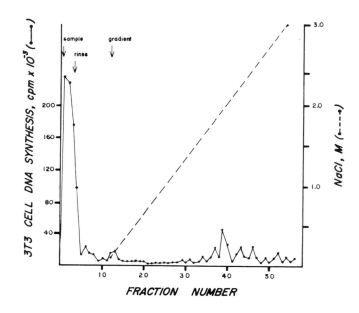

Figure 9:   Chondroitin sulfate-Sepharose chromatography of RDGF.   A
crude extract of retinas was prepared as outlined in the legend of
Figure 6.   The column was eluted with a gradient of 0.1-2.0 M NaCl
and the fractions assayed for EC and 3T3 cell stimulatory activity.
None of the stimulatory activity that was previously shown to bind
with high affinity to heparin binds to chondroitin sulfate.

suggesting that the interaction between RDGF and heparin is a
specific one.

Further evidence for the specificity of the growth factor-
heparin interaction is found by comparison of the heparin binding
ability of RDGF to another similarly charged growth factor.   EGF, is
a polypeptide growth factor with an isoelectric point of approximately 5. If the binding between RDGF and heparin is based solely on
charge interactions, then EGF should bind similarly to a heparin
affinity column.   However, as the data in Figure 10 illustrate, EGF
does not bind to heparin.   This finding indicates that net anionic
charge is not the basis for the binding of RDGF to heparin and
confirms the specific nature of the heparin-RDGF affinity.

The third piece of evidence for the specificity of the heparin-
growth factor interaction is demonstrated in studies that examine
the effect of the GAGs on the mitogenic action of RDGF.   Addition of
heparin or heparan sulfate and a subsaturating concentration of RDGF
to cultures of capillary EC potentiates the mitogenic activity of

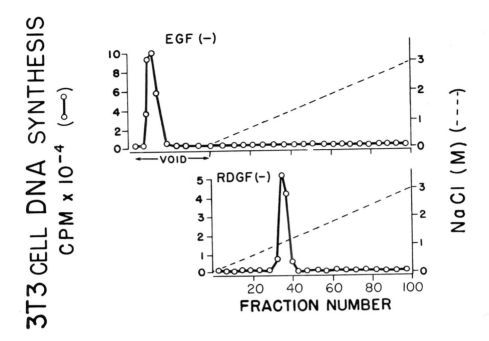

Figure 10: Comparison of the behavior of epidermal growth factor (EGF) and RDGF on a heparin affinity column. The growth factors were prepared and the columns run as outlined in the legend of Figure 6. In contrast to RDGF, EGF, another anionic polypeptide growth factor, does not bind at all to immobilized heparin.

the growth factor (Figure 11) (Orlidge & D'Amore, 1986). In contrast, neither the chondroitin sulfates nor hyaluronic acid potentiates the effect of RDGF (Figure 12), providing further support for the idea that RDGF, heparin-like molecules, and the endothelium interact specifically.

SUMMARY

The results of our studies show that the retina contains an anionic polypeptide growth factor, RDGF, which binds with high affinity to heparin but not to chondroitin sulfate, another sulfated GAG. RDGF stimulates the proliferation of both large and small vessel EC and elicits new blood vessel growth in vivo. Investigation of the action of GAGs on vascular cells in vitro reveals that heparin and HA have cell-specific effects on their

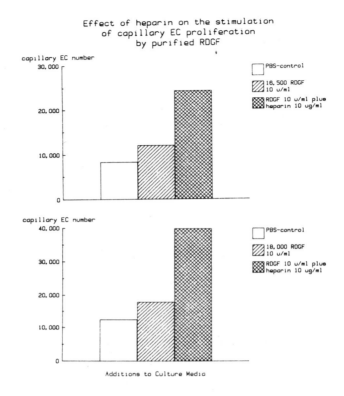

Figure 11: EC stimulatory activity of purified RDGF and its potentiation by heparin. Purified RDGF exists in two forms with molecular weights of 16,500 and 18,000. Both forms maximally stimulate the proliferation of EC at 20 units/ml (data not shown). Addition of 10 μg/ml of heparin to a subsaturating dose (10 units/ml) of RDGF results in potentiation of its mitogenic activity.

attachment and proliferation. HA inhibits the attachment of the pericytes and enhances the attachment of the EC. Heparin inhibits the proliferation of pericytes and potentiates the growth of EC.

One possible scenario for the interaction of RDGF and the matrix-associated heparin-like molecules in the control of vessel growth in the eye is presented in the model in Figure 13. When a capillary is mature and fully differentiated it is surrounded by a basement membrane in which heparin-like molecules are the predominant glycosaminoglycans. These basement membrane-associated heparin-like molecules may act as solid-phase binding sites,

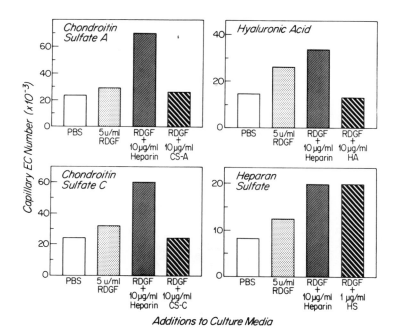

Figure 12: Effect of GAG addition on the mitogenic activity of RDGF. The addition of 10 µg/ml of heparin potentiates the mitogenic activity of a suboptimal dose of RDGF. Heparan sulfate is also effective in the potentiation of growth factor at a 10-fold lower concentration. Neither the chondroitin sulfates nor hyaluronic acid potentiated the effect of RDGF, indicating the specificity of the interaction between RDGF and heparin-like molecules.

immobilizing RDGF and making it inaccessible to the EC. In con-trast, during the process of capillary formation, the basement mem-brane is incomplete and HA is the predominant glycosaminoglycan species. In this case, heparin-binding mitogens such as RDGF would not be immobilized but rather would be free to interact and stimulate the EC. The pericyte could be similarly influenced by changes in the profile of glycosaminoglycans that are present. In the mature capillary, the presence of heparin-like components in the basement membrane might act to suppress the proliferation of the pericytes, helping to maintain them as a single layer. The association of the pericyte with the capillary EC is an event

MATURE CAPILLARY

IMMATURE CAPILLARY

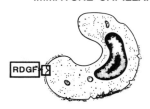

Figure 13: Proposed model for the interaction of RDGF, heparin and endothelial cells.

thought to mark the maturation of the capillary. The high levels of HA known to characterize the immature capillary may be responsible for keeping the capillary free from pericytes until formation is completed. Thus, the action of RDGF together with the influences of the matrix-and cell-associated glycosaminoglycans may act in concert to regulate the formation of new blood vessels in the eye.

REFERENCES

Abatangelo, G., R. Cortivo, M. Martelli and P. Vecchia, 1982. Cell detachment mediated by hyaluronic acid. Exp. Cell. Res., 137:73-78.

Abatanglo, G., M. Martelli and P. Vecchia, 1983. Healing of hyaluronic acid-enriched wounds: histological observations. J. Surg. Res., 35:410-416.

Ausprunk, D. H. and J. Folkman, 1977. Migration and proliferation of endothelial cells in preformed and newly formed blood vessels during tumor angiogenesis. Microvasc. Res., 14:53-65.

Ausprunk, D.H., 1982. Synthesis of glycoproteins by endothelial cells in embryonic blood vessels. Dev. Biol., 90:79-90.

Ashton, N., 1961. Neovascularization in ocular disease. Trans. Ophthalmol. Soc. U.K., 81:145.

Barnhart, B.J., S.H. Cox and P.M. Kraemer, 1979. Detachment variants of Chinese hamster cells. Hyaluronic acid as a modulator of cell detachment. Exp. Cell Res., 119:327-332.

Castellot, J.J.Jr., M.L. Addonizio, R.D. Rosenberg and M.J. Karnovsky, 1981. Cultured endothelial cells produce a heparin-like inhibitor of smooth muscle cell growth. J. Cell Biol., 90:372-379.

Castellot, J.J. Jr., L.V. Favreau, M.J. Karnovsky and R.D. Rosenberg, 1982. Inhibition of vascular smooth muscle cell growth by endothelial cell-derived heparin. Possible role of a platelet endoglycosidase. J. Biol. Chem., 257:11256-11260.

Clowes, A.W. and M.J. Karnovsky, 1977. Suppression by heparin of smooth muscle cell proliferation in injured arteries. Nature (London), 265:625-626.

Cole, G.J., D. Schubert and L. Glaser, 1985. Cell substratum adhesion in chick neural retina depends upon protein-heparan sulfate interactions. J. Cell Biol., 100:1192-1199.

Courtoy, P.J. and J. Boyles, 1983. Fibronectin in the microvasculature: localization in the pericyte-endothelial interstitium. J. Ultrastruc. Res., 83:258-273.

Crocker, D.J., T.M. Murad and J.C. Geer, 1970. Role of the pericyte in wound healing: an ultrastructural study. Exptl. Molec. Path., 13: 51-65.

Culp, L.A., B.A. Murray and B.J. Rollins, 1979. Fibronectin and proteoglycans as determinants of cell-substratum adhesion. J. Supramol. Struc., 11:401-427.

D'Amore, P.A., B.M. Glaser, S.K. Brunson and A.H. Fenselau, 1981. Angiogenic activity from bovine retina: Partial purification and characterization. Proc. Natl. Acad. Sci. USA, 78:3068-3072.

D'Amore P.A. and M. Klagsbrun, 1984. Endothelial cell mitogens derived from retina and hypothalmus: Biochemical and biological similarities. J. Cell Biol., 99:1545-1549.

Engerman, R.L., D. Pfaffenbach and M.D. Davis, 1967. Cell turnover of capillaries. Lab. Invest., 17:738-743.

Fisher, M. and M. Solursh, 1979. Influence of substratum on mesenchyme spreading in vitro. Exp. Cell Res., 123:1-14.

Forrester, J.V. and P.C. Wilkinson, 1981. Inhibition of leukocyte locomotion by hyaluronic acid. J. Cell Sci., 48:315-331.

Gimbrone, M.A. Jr., R.S. Cotran, S.B. Leapman and J. Folkman, 1974. Tumor growth and neovascularization: An experimental model using the rabbit cornea. J. Natl. Cancer Instit., 52:413-427.

Glaser, B.M., P.A. D'Amore, R.G. Michels, A. Patz and A. Fenselau, 1980. Demonstration of vasoproliferative activity from mammalian retina. J. Cell Biol., 84:298-304.

Gospodarowicz, D., D. Delgado and I. Vlodavsky, 1980. Permissive effect of the extracellular matrix on cell proliferation in vitro. Proc. Natl. Acad. Sci. USA, 77:4094-4098.

Gospodarowicz, D., G. Greenberg, and C.R. Birdwell. 1978. Determination of cellular shape by the extracellular matrix and its correlation with control of cellular growth. Cancer Res., 38:4155-4171.

Grinnell, F., M.K. Feld and K. Minter, 1980. Fibroblast adhesion to fibrinogen and fibrin substrata: requirement for cold insoluble globulin. Cell, 19:517-525.

Hay, E.D., 1981. Extracellular matrix. J. Cell Biol., 91 No.3, pt. 2:2055-2235.

Hay, E.D., 1985. Matrix-cytoskeletal interactions in the developing eye. J. Cellular Biochim. Biophys. Acta., 27:143-156.

Kawakami, H. and H. Terayama, 1981. Liver plasma membrane and proteoglycans prepared there from inhibit the growth of hepatoma cells in vitro. Biochim Biophys. Acta., 646:161-168.

Kleinman, H.K., R.J. Klebe and G.R. Martin, 1981. Role of collagenous matrices in the adhesion and growth of cells. J. Cell Biol., 88:473-485.

Kraemer, P.M. and R.A. Tobey, 1972. Cell cycle-dependent desquamation of heparan sulfate from the cell surface. J. Cell Biol., 55:713-717.

Kujawa, M.J. and K. Tepperman, 1983. Culturing chick muscle cells on glycosaminoglycan substrates: attachment and differentiation. Dev. Biol., 99:277-286.

Lander, A.D., D.K. Fujii, D. Gospodarowicz and L.F. Reichardt, 1982. Characterization of a factor that promotes neurite outgrowth: evidence linking the activity to heparan sulfate proteoglycan. J. Cell Biol., 94:574-585.

Langer, R., 1981. Polymers for sustained release of macromolecules. Van Vunakis, H. and J.J. Langonne, eds., Immunological Techniques. In: Methods in Enzymology, 73:57-75.

Lindahl, U. and M. Hook, 1978. Glycosaminoglycans and their binding to biological macromolecules. Ann. Rev. Biochem., 47:385-417.

Madri, J.A. and S.K. Williams, 1983. Capillary endothelial cell cultures: phenotypic modulation by matrix components. J. Cell Biol., 97:153-165.

Majack, R.A. and A.W. Clowes, 1984. Inhibition of vascular smooth muscle cell migration by heparin-like glycosaminoglycans. J. Cell Physiol., 118:253-256.

Martinez-Hernandez, A., and P.S. Amenta. 1983. The basement membrane in pathology. Lab. Invest., 48:656-677.

Matuoka, K. and Y. Mitsui, 1981. Involvement of cell surface heparan sulfate in the density-dependent inhibition of cell proliferation. Cell Struct. Funct., 6:23-33.

Murray, J.C., G.S. Stingl, H.K. Kleinman, G.R. Martin and S.I. Katz, 1979. Epidermal cells adhere preferentially to type IV (basement membrane) collagen. J. Cell Biol., 80:197-202.

Orlidge, A. and P.A. D'Amore, 1986. Cell specific effects of glycosaminoglycans on the attachment and proliferation of vascular wall components. Microvas. Res. 31:41-53.

Pratt, B.M., A.S. Harris. J.S. Morrow and J.A. Madri, 1984. Mechanisms of cytoskeletal regulation. Modulation of aortic endothelial cell spectrin by the extracellular matrix. Am. J. Pathol., 117:337-342.

Sage, H. and P. Bornstein. 1982. Endothelium from umbilical vein and hemangioendothelioma secrete basement membrane largely to the exclusion of interstitial procollagens. Arteriosclerosis, 2:27-36.

Schubert, D. and M. LaCorbiere, 1980. Role of a 16S glycoprotein complex in cellular adhesion. Proc. Natl. Acad. Sci., USA, 77:4137-4141.

Shing, Y., J. Folkman, R. Sullivan, C. Butterfield, J. Murray and M. Klagsbrun, 1984. Heparin affinity: Purification of a tumor-derived capillary endothlelial cell growth factor. Science, 223:1296-1298.

Stamatoglou, S. and J.M. Keller, 1983. Correlation between cell substrate attachment in vitro and cell surface heparan sulfate affinity for fibronectin and collagen. J. Cell Biol., 96:1820-1823.

Terranova, V.P., D.H. Rohrbach and G.R. Martin, 1980. Role of laminin in the attachment of Pam 212 (epithelial) cells to basement membrane collagen. Cell. 22:719-726.

Tomida, M., H. Koyama and T. Ono, 1974. Hyaluronic acid synthetase in cultured mammalian cells producing hyaluronic acid. Oscillating changes during the growth phase and suppression by 5-bromodeoxyuridine. Biochim Biophys. Acta., 338:352-363.

Toole, B.P., 1981. Glycosaminoglycans in morphogenesis. In "Cell Biology of the Extracellular Matrix." E.D. Hay, ed., pp. 259-294, Plenum Press, NY.

Toole, B.P. and J. Gross, 1971. The extracellular matrix of the regenerating newt limb: synthesis and removal of hyaluronate prior to differentiation. Dev. Biol., 25:57-77.

Underhill, C.B. and J.M. Keller, 1976. Density-dependent changes in the amount of sulfated glycosaminoglycans associated with mouse 3T3 cells. J. Cell. Physiol., 89:53-63.

Wise, G.N. 1956. Retinal neovascularization. Trans. Am. Ophthalmol. Soc., 54:729-826.

Yamada, K.M. and K. Olden, 1978. Fibronectins--adhesive glycoproteins of cell surface and blood. Nature (London), 275:179-184.

Yamada, K.M., S.K. Akiyama and M. Hayashi, 1980a. Fibronectin structure and function and its interaction with glycosaminoglycans. Biochem. Soc. Trans., 9:506-508.

Yamada, K.M., D.W. Kennedy, K. Kimata and R.M. Pratt, 1980b. Characterization of fibronectin interactions with glycosamino-glycans and identification of active proteolytic fragments. J. Biol. Chem., 255:6055-6063.

Yamagami, I., 1970. Electron microscopic study on the cornea. I. The mechanism of experimental new vessel formation. Japan J. Ophthalmol., 14:41-58.

Young, W.C. and I.M. Herman, 1985. Extracellular matrix modulation of endothelial cell shape and motility following injury in vitro. J. Cell Sci., 73:19-32.

Elucidation of the Visual Cycle by Study of Two Retinoid-Binding Proteins

John C. Saari, Ann H. Bunt-Milam, and Vinod Gaur

The visual cycle (Wald, 1968) includes all metabolic reactions of vitamin A known to occur in retina, with the exception of those that generate retinoic acid, a retinoid with no known role in the visual process. A modern version of the visual cycle, shown in Figure 1, is very similar to the cycle outlined by Wald in 1968 except for a few details. Many of the questions of interest at that time remain with us today. The nature of the generation of the 11-cis-configuration in the dark remains unclear. Is the process enzymatic? Is it the aldehyde, alcohol or retinyl ester that is isomerized? In which cell type(s) of the retina does the isomerization take place? How many dehydrogenases function in the visual cycle? How are these reactions controlled? How do retinoids traverse the extracellular space that separates the retinal pigment epithelium (RPE) and neural retina? The answers to these and other questions are not yet known, partly because of the difficulties inherent in dealing with water insoluble substrates and membrane associated enzymes. Until recently, analytical methods available for retinoid separations, especially separation of geometrical isomers, were tedious and not well suited to the task. Currently there is renewed interest in the reactions of the visual cycle, stimulated in part by development of rapid and efficient methods for the separation of retinoids by HPLC. In addition, several water soluble, retinoid-binding proteins have been discovered in retina and other tissues (Adler and Martin, 1982; Futterman et al., 1977; Liou et al., 1982; Wiggert and Chader, 1975). There is a growing consensus among investigators that these proteins participate in the visual cycle, perhaps as substrate carrier proteins for enzymatic reactions and/or as transport vehicles for intracellular and extra-

THE VISUAL CYCLE

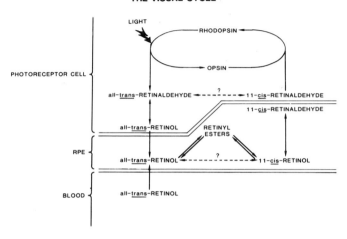

Figure 1. Metabolism of vitamin A in retina. The reactions that are known to occur in neural retina and RPE are indicated. The location and nature of the isomerization process are not known at this time.

cellular diffusion of retinoids.   This chapter will attempt to summarize the influence that study of two of these proteins has had on our understanding of the visual cycle. Although the emphasis, of necessity, will be on work originating in the authors' laboratories, it should be noted that a number of other laboratories have made significant contributions in this area (reviewed in Bridges, 1984; Chader, 1982).

The visual cycle begins with the photoisomerization of 11-cis-retinaldehyde bound to opsin to all- trans-retinaldehyde, the event that triggers the electrophysiological response of the photoreceptor cell to light.   All-trans-retinaldehyde is reduced to all-trans-retinol by retinol dehydrogenase in the outer segment (Wald and Hubbard, 1949).   This membrane bound enzyme is coupled to pentose cycle activity in the photoreceptor cell inner segment, due to its requirement for NADPH (Futterman, 1963). Although reports persist of the ability of soluble alcohol dehydrogenase to reduce retinaldehyde

(Julia et al., 1983), it seems unlikely that this enzyme plays a role in retinoid metabolism. After prolonged bleaching, all-trans-retinol disappears from neural retina and appears in the RPE as retinyl ester (Bridges, 1976; Dowling, 1960). Little is known about translocation of retinol across the plasma membranes of the outer segment and RPE and through the extracellular compartment (interphotoreceptor space) that separates them. Recent discovery and characterization of a retinoid-binding protein (interphotoreceptor retinoid-binding protein or interstitial retinol-binding protein, IRBP) in this space is consistent with protein mediated transport of retinol (Adler and Martin, 1982; Bunt-Milam and Saari, 1983, Lai et al., 1982; Liou et al., 1982). Retinyl ester synthase activity has been reported in homogenates of neural retina and RPE (Andrews and Futterman, 1963; Krinsky, 1958). However, the activity in neural retina is less than 1% that of RPE (Berman et al., 1980) and probably results from contamination from RPE cells. Apparently a single enzyme is responsible for the synthesis of both 11-cis- and all-trans-retinyl esters. Retinyl ester hydrolase activity has been reported in the particulate fractions of RPE (Berman et al., 1982), although it has never been clearly established that the enzyme is specific for retinyl ester. The maximum activity reported is much less than that of retinyl ester synthase, raising questions about the control and coordination of the two opposing enzymatic processes. RPE microsomes also contain a dehydrogenase specific for cis-retinols, distinct from the all-trans-retinol dehydrogenase of outer segments (Lion et al., 1975; Zimmerman et al., 1975). A similar enzyme has also been found in neural retina of frog and toad, although the mammalian retinas tested to date do not show this dehydrogenase activity (Perlman et al., 1982; Yoshikami and Noll, 1978). All-trans-retinol in RPE is primarily derived from the blood via receptor-mediated uptake (Bok and Heller, 1976). Regeneration of the 11-cis-configuration in the dark remains unexplained. Earlier reports of an isomerase indicated that the enzyme lacked sufficient activity in the dark to provide 11-cis-retinaldehyde at the rate required for regeneration of visual pigment (Hubbard, 1956). Other reports of isomerase activity were later shown to result in generation of only 9- or 13-cis- but not 11-cis-retinaldehyde (Futterman, 1974; Ostapenko and Furayev, 1973; Rotmans et al, 1972). At present it is unclear if the reaction is enzymatic (Bernstein et al., 1985), if it is localized to a specific cell type, or which form of the vitamin is isomerized.

DARK ISOMERIZATION OF ALL-TRANS-RETINALDEHYDE MAY NOT BE ENZYMATIC

It is ironic that we know the least about the most fundamental dark-reaction of the visual cycle. Following a bleach, regeneration of visual pigment in the dark depends on the production of 11-cis-retinaldehyde. The process in humans occurs with a time constant of ca. 7 minutes (Alpern, 1971) and appears to be stereospecific, in that only the 11-cis-isomer of retinaldehyde is found in visual pigments. This strict stereospecificity has led most investigators to assume that the isomerization must be enzymatic. Indeed, one frequently finds the terms "retinal or retinol isomerase" in reviews and textbooks of biochemistry (Lehninger, 1982; Vance, 1983). However, failure of a number of laboratories to confirm the enzymatic nature of this reaction plus more recent experimental evidence (Bernstein et al., 1985) lead to the tentative conclusion that the reaction may not be enzymatic. Dark, non-enzymatic isomerizations of all-trans-retinaldehyde produce an equilibrium mixture of geometrical isomers including 0.1% 11-cis-retinaldehyde (Rando and Chang, 1983). If this is the process for the generation of the 11-cis-configuration for visual pigment regeneration, then stereospecificity must be introduced into the system at some other step. Opsin forms complexes in vitro with either 9- or 11-cis-retinaldehyde; thus the stereospecificity must be introduced before the retinoid is presented to the bleached visual pigment and perhaps after isomerization. Hubbard, one of the original investigators in this field, recognized that stereospecifity could be introduced by a binding protein specific for the 11-cis-configuration. In a paper with Colman, she speculated on the possibility of trapping 11-cis-retinaldehyde from a mixture of different configurations. "Such trapping could involve the binding of 11-cis-vitamin A by specific proteins or its selective esterification" (Hubbard and Colman, 1959). In the following section, the properties of a protein similar to the one hypothesized by Hubbard and Colman will be described.

Four retinoid-binding proteins have been demonstrated in extracts of neural retina (reviewed in Bridges, 1984; Chader, 1982; Chytil and Ong, 1982). Three of these proteins form complexes with specific classes or configurations of retinoids. Cellular retinoic acid-binding protein (CRABP) binds only retinoic acids, cellular retinol-binding protein (CRBP) binds only retinols and cellular retinaldehyde-binding protein (CRALBP) binds only retinols and retinaldehydes of the cis-configuration (Futterman et al., 1977). In contrast, IRBP binds many hydrophobic substances, including retinoids (Fong et al., 1984).

CRALBP was originally discovered because of its ability to form stable, non-covalent complexes with exogenous 11-cis-retinaldehyde. Subsequent purification of the protein-retinoid complex was accomplished by following the radioactivity of bound, tritiated retinoid. The protein is a water soluble, acidic molecule with a molecular weight of 33,000, approximately twice the molecular weight of CRBP and CRABP (Stubbs et al., 1979). The presence of a blocked amino terminus has prevented sequence analysis to date (Crabb and Saari, 1981). Therefore, the relationship of this protein to the family of sequence-related proteins that includes CRBP and CRABP (Sundelin et al., 1985) must await more detailed structural studies. Relatively large amounts of CRALBP are present in bovine neural retina (1 nmole per eye) and retinal pigment epithelium (0.7 nmole per eye). CRALBP has not been detected in other tissues of the rat, suggesting that it is unique to primary visual tissues (Futterman and Saari, 1978).

It was important to demonstrate that CRALBP co-purified with endogenous retinoids, since many proteins not related to vitamin A metabolism can form complexes with exogenous vitamin A (Saari et al., 1982). Figure 2 illustrates the absorption spectrum for CRALBP obtained from bovine RPE in the absence of exogenous ligands. The smooth absorption maximum centered at 425 nm is due to the presence of bound 11-cis-retinaldehyde, as shown by extraction of the retinoids from the protein and determination of their composition by reverse phase HPLC (figure 3). Approximately 1 mole of 11-cis-retinaldehyde is found per mole of CRALBP, showing that the binding site is fully occupied by endogenous retinoid. Isolation of CRALBP with its endogenous ligand from neural retina produced a surprising result. The absorption spectrum of the protein, (figure 2) has two maxima in addition to the usual protein absorption at 280 nm; one at 425 nm, due to bound 11-cis-retinaldehyde, and a second at 330 nm, more typical of a retinol. Indeed, HPLC analysis of the extracted endogenous retinoids reveals the presence of 11-cis-retinaldehyde and 11-cis-retinol (figure 3). Thus, CRALBP isolated

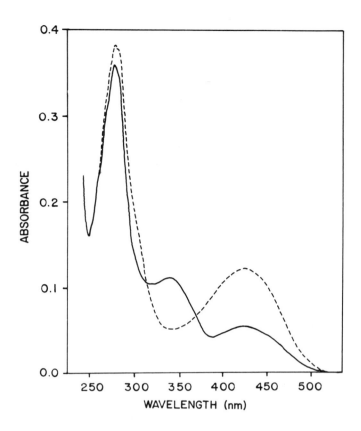

Figure 2.　Absorption spectra of bovine CRALBP purified with bound, endogenous retinoids. (——) CRALBP from neural retina. (– – –) CRALBP from RPE.

from neural retina is complexed with two ligands of the 11-cis-configuration. Since RPE apical processes adhere to the retina during its mechanical separation from RPE (Bunt-Milam and Saari, 1983), it is possible that CRALBP carrying 11-cis-retinaldehyde in neural retinal extracts may actually be derived from RPE. None-the-less, the results strongly reinforce the contention that CRALBP is involved in the visual cycle, since it carries endogenous retinoid with the 11-cis-configuration that, to date, is implicated solely in the visual process.

As noted above, interaction of 11-cis-retinaldehyde with CRALBP produces a complex with a chromophoric absorption maximum of 425 nm,

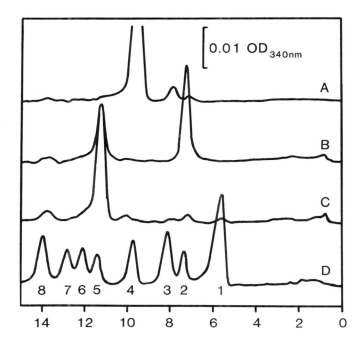

Figure 3. Reverse phase HPLC analysis of the endogenous retinoids extracted from bovine retinoid-binding proteins. A. CRBP from neural retina. B. CRALBP from neural retina. C. CRALBP from RPE. D. Standards: 1, all-trans-retinoic acid; 2, 11-cis-retinol; 3, 9- and 13-cis-retinol; 4, all-trans-retinol; 5, 11-cis-retinaldehyde; 6, 13-cis-retinaldehyde; 7, 9-cis-retinaldehyde; 8, all-trans-retinaldehyde.

shifted well to the red of the 365 nm absorption maximum of 11-cis-retinaldehyde in solution. Such a bathochromic shift is characteristic of visual pigments such as rhodopsin. From this point of view, CRALBP carrying 11-cis-retinaldehyde resembles a blue sensitive visual pigment in its spectral properties and might be expected to show photosensitivity typical of a visual pigment. When a solution of CRALBP 11-cis-retinaldehyde is illuminated, a striking alteration is noted in the chromophoric absorption peak (figure 4) (Saari et al., 1984). The maximum at 425 nm disappears and is replaced by a stronger absorbance centered at 365 nm. HPLC analysis of the extracted retinoids reveals that illumination causes a photochemical isomerization of 11-cis-retinaldehyde to all-trans-retinaldehyde with the generation of minor amounts of 13-cis-retinaldehyde (figure 5). The spectral shifts result from the

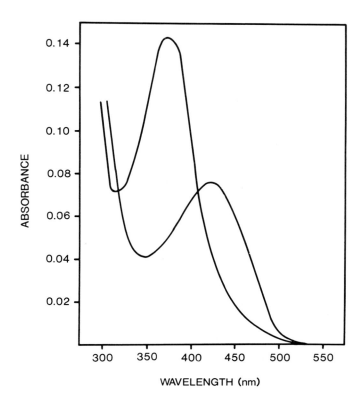

Figure 4. Absorption spectra of CRALBP 11-cis-retinaldehyde complex. Right peak, CRALBP maintained in red light. Left peak, CRALBP illuminated with white light for 180 seconds.

disappearance of the CRALBP·11-cis-retinaldehyde complex with its bathochromically shifted absorption maximum and low extinction coefficient and the generation of all-trans-retinaldehyde, which has no affinity for CRALBP and thus displays its solution absorption maximum and higher extinction coefficient.

The photosensitivity of CRALBP·11-cis-retinaldehyde _in_ _vitro_ led us to determine the extent of photoisomerization in retina of 11-cis-retinaldehyde bound to CRALBP compared to that bound to opsin. Surprisingly, the complex is quite photostable. Illumination that bleaches approximately 70% of the rhodopsin does not result in demonstrable photoisomerization of 11-cis-retinaldehyde bound to CRALBP. Photosensitivity is defined as the product of the extinction

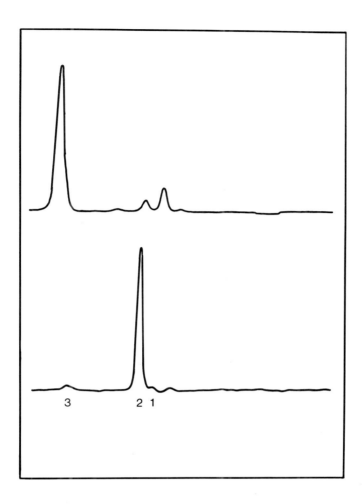

Figure 5. Retinoids extracted from CRALBP of figure 4 and analyzed by normal phase HPLC. Lower curve, CRALBP maintained in red illumination. Upper curve, CRALBP illuminated with white light for 180 seconds. 1, 13-cis-retinaldehdye; 2, 11-cis-retinaldehyde; 3, all-trans-retinaldehyde.

coefficient (ε_M) and the quantum efficiency for photoisomerization (Φ_{PI}) (Dartnall, 1972). The ε_M for CRALBP.11-cis-retinaldehyde at 425 nm (15,400 M^{-1} cm^{-1}) is about one third that of rhodopsin at 500 nm. However, this 3- fold difference is not sufficient to explain the undetectable bleaching of CRALBP 11-cis-retinaldehyde we observed. Thus it seemed likely that Φ_{PI} also had to be low to explain our results. An estimate of Φ_{PI} relative to the known value for rhodopsin (Φ_{PI}= 0.69, Dartnall, 1972) can be obtained by comparing the rates of photoisomerization of the two complexes in vitro. The results indicate that Φ_{PI}<0.1. Thus, the low photosensitivity of the complex is due to both a low extinction coefficient and a low quantum yield for photoisomerization. It seems unlikely that the physiologic function of CRALBP is related to its superficial resemblance to a visual pigment but is more probably associated with the low photosensitivity of the complex. In other words, CRALBP may function to stabilize 11-cis-retinaldehyde rather than to facilitate its photoisomerization as postulated for opsin.

Early experiments that examined the binding specificity of CRALBP were performed using a competition assay in which the ability of a retinoid in 100 fold molar excess to displace bound 11-cis-retinaldehyde was scored (Futterman et al., 1977). These experiments, performed with retinal extracts, indicated that retinaldehydes of the cis-configuration were most effective in displacing 11-cis-retinaldehyde and that neither retinols nor retinoic acids were effective. More recently, we have examined this question using purified CRALBP and HPLC-purified retinoids. The results indicate that the binding site is selective, in that only 9-cis- and 11-cis-retinaldehyde can form complexes with CRALBP, but gives no indication of the relative affinities of the ligands for the binding protein. When CRALBP is mixed with an equimolar mixture of 9-, 11-, 13-cis- and all-trans-retinaldehydes it becomes apparent that the association constant (K_a) must be very much greater for 11-cis-retinaldehyde than those for the other aldehydes, since CRALBP isolated from this mixture is associated only with 11-cis-retinaldehyde. Thus, it seems that the binding site of apo-CRALBP is sufficiently restrictive to confer the stereochemical specificity required of the visual system.

A recent report by Rando and associates suggests that the dark isomerization of all-trans-retinaldehyde may not be enzymatic (Bernstein et al., 1985). If a non-stereospecific, non-enzymatic catalyst could generate an equilibrium mixture of geometrical isomers of retinaldehyde, apo-CRALBP could provide the needed stereospecificity by virture of its high affinity for 11-cis-retinaldehyde. As much 11-cis-retinaldehyde could be formed as there is apo-CRALBP to capture it unless mechanisms exist to transfer it to a sink such as opsin or to sequester it in some way, perhaps as retinyl ester. Thus, there is currently no evidence to support the classic model of stereospecific enzymatic regeneration of the 11-cis-configuration in the dark.

PROTEIN-BOUND RETINOIDS CONSTITUTE POOLS OF SUBSTRATES FOR VISUAL CYCLE ENZYMES

Protein-bound pools of retinol exist in neural retina and RPE (Saari et al., 1982). In retina both all-trans- and 11-cis-retinol are present, the former bound to CRBP and and latter bound to CRALBP (figure 3). This represents an interesting example of intracellular compartmentalization, since CRBP forms stable complexes with 11-cis-retinol in vitro (figure 6), yet carries none of it in vivo. All the 11-cis-retinol is carried by CRALBP. This specificity is probably conferred during the loading process or perhaps by the affinity of the binding protein for retinols. In RPE there is a pool of water soluble, all-trans-retinol bound to CRBP. CRALBP from RPE has only 11-cis-retinaldehyde associated with it. 11-cis-retinol is associated only with the protein derived from neural retina (Saari et al., 1982). Are these protein-bound pools of retinol metabolically active? A direct answer to that question has not yet been obtained. However, protein-bound retinols do serve as substrates for enzymes that modify retinol. The only known visual cycle enzyme in neural retina is all-trans-retinaldehyde dehydrogenase found in photoreceptor cell outer segments. However, RPE is a known source of three enzymes of the visual cycle. The microsomal fraction from RPE, for instance, is an excellent source of two enzymes that metabolize retinol, 11-cis-retinaldehyde dehydrogenase and retinyl ester synthase (Berman et al., 1980; Zimmerman et al., 1975). The latter term is used here provisionally to describe the enzymatic synthesis of retinyl ester that occurs when retinol is added to RPE microsomes. It is possible to study one of these reactions without competition from the other because of the difference in pH optima of the two reactions and the requirement for pyridine nucleotide by the dehydrogenase reaction.

Berman and co-workers were the first to demonstrate that retinol bound to CRBP is a substrate for the retinyl ester synthase of RPE cells (Berman et al., 1980). Studies in our laboratory have confirmed their findings (figure 7) and provide further evidence that the product of the reaction, retinyl ester, is removed from the

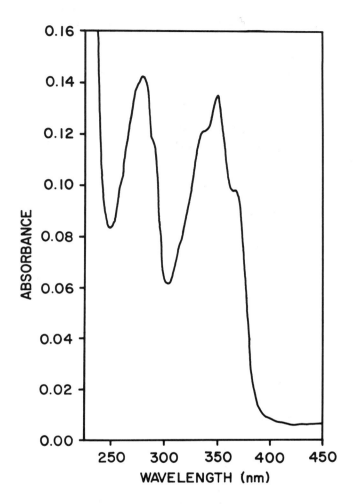

Figure 6. Absorption spectrum of CRBP complexed with 11-cis-retinol. HPLC analysis of the complex revealed 92% 11-cis-retinol, 8% all-trans-retinol.

binding protein and becomes associated with the microsomes (Saari et al., 1984). This reaction is the first described in the visual system for the removal of a bound retinoid from CRBP. The Ka of CRBP for all-trans-retinol is reported to be 3 X 10^8 M^{-1} corresponding to ΔG^0 = -9.5 kcal/mol (Wiggert et al., 1977).

11-cis-retinol bound to CRALBP is a better substrate than all-trans-retinol bound to CRBP for the retinyl ester synthase by RPE

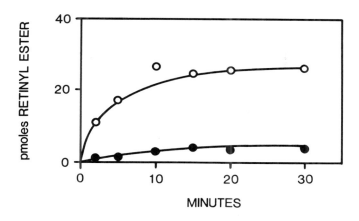

Figure 7. Esterification of all-trans-retinol complexed with CRBP (●) and 11-cis-retinol complexed with CRALBP (O) by microsomes from bovine RPE at O.

microsomes (figure 7) and again the product, 11-cis-retinyl ester, is removed from the binding protein during the reaction. Does this reflect a different metabolic routing of the retinols bound to CRBP and CRALBP? Our finding that CRBP from RPE is saturated with all-trans-retinol whereas CRALBP has no 11-cis-retinol bound to it is consistent with this suggestion (Saari et al., 1982). Any 11-cis-retinol bound to CRALBP may have been esterified and "stripped" from the binding protein.

Although the ability of protein-bound retinols to serve as substrates for the retinyl ester synthase suggests a physiological role, the enzyme also readily esterifies retinol added to microsomes in the absence of binding protein. Thus, there is no actual evidence that esterification of the protein-bound component is of physiological importance, although the suggestion seems plausible.

11-cis-retinol dehydrogenase of RPE microsomes also reduces free and protein-bound 11-cis-retinaldehyde with approximately equal efficiency. However, the interaction of this dehydrogenase with 11-cis-retinaldehyde bound to CRALBP suggests there is recognition between the binding protein and dehydrogenase or a microsomal component (Saari and Bredberg, 1982). Whereas microsomal dehydrogenase will readily reduce 11-cis-retinaldehyde bound to CRALBP, the aldehyde functional group is not reactive with low molecular weight chemical reducing agents or with NH_2OH. The possibility that a component of the microsomes induces the formation of a different CRALBP-retinoid conformation that exposes the aldehyde group seems unlikely since incubation of the complex with $NaBH_4$ in the presence of microsomes without NADH does not allow reduction. This striking feature of the dehydrogenase reaction demonstrates that the enzymatic reduction of retinaldehyde bound to CRALBP is more than fortuitous interaction of two proteins. However, the demonstration of a physiologic role for this process awaits further experimentation.

In summary, two visual cycle enzymes exist in RPE that use protein-bound retinoids as substrates. A hypothetical coupling of these two enzymes involving CRALBP is shown in figure 8. The role of protein-bound retinoids as substrates in neural retina is less clear since the only known enzymatic reaction of retinol in this tissue takes place in a compartment (the outer segment) that appears to be free of retinoid-binding proteins (Bok et al., 1985; Bunt-Milam and Saari, 1984).

MULLER GLIAL CELLS MAY BE INVOLVED IN THE VISUAL CYCLE

The availability of highly purified CRALBP has enabled production of antibodies to the protein that can be used to determine the cellular localization of CRALBP within the retina (Bunt-Milam and Saari, 1983; Bunt-Milam et al., 1985). Fortunately, CRALBP is an excellent antigen in rabbits, in contrast to CRABP and CRBP. The specificity of polyclonal anti-CRALBP produced in rabbits has been determined by separating retinal homogenate proteins by size (SDS-PAGE) or charge (non-SDS-PAGE), transferring the proteins to nitrocellulose (Western blotting) and indirect immunostaining using the PAP procedure. Of the hundreds of proteins present, anti-CRALBP recognizes a single protein with size and charge identical to those of CRALBP. Localization studies performed using indirect immunofluorescence reveals a striking pattern of staining (see figure 9 for an example with a monoclonal antibody). The RPE is strongly fluorescent relative to the control. This result confirms earlier biochemical studies showing CRALBP in extracts of bovine RPE. The pattern of anti-CRALBP immunofluorescence in neural retina is more surprising. CRALBP is known to be present in extracts of neural retina and thus staining with anti-CRALBP was anticipated.

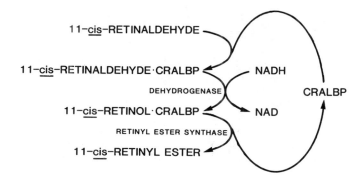

Figure 8. Hypothetical scheme for the coupling of two visual cycle
enzymes of RPE. Coupling of the two enzymatic reactions with
substrates complexed with CRALBP has been demonstrated in vitro.
However, as yet there is no evidence that the pathway functions in
vivo.

In view of the metabolism of vitamin A and the visual pigment in
photoreceptor cells, we expected that CRALBP would be found in
photoreceptors, possibly the inner segments. However, the
photoreceptor inner and outer segments are consistently free of
anti-CRALBP immuno-fluorescence. The specific labeling spans the
entire retina from the inner to the outer limiting membrane, and
includes foot processes against the inner limiting membranes, somata
in the inner nuclear layer, fine processes ramifying in the inner
and outer plexiform layers and microvilli extending from the outer
limiting membrane into the interphotoreceptor space. The retinal
cell that corresponds to this staining pattern is the Muller cell,
the major glial cell of the retina (Sarthy and Bunt, 1982; Uga and
Smelser, 1973). We have recently confirmed this unexpected result of
Muller cell labeling with a monoclonal antibody to CRALBP (figure
9). Characterization of this antibody by immunoprecipitation
demonstrates it to be specific for CRALBP (figure 10). Electron
microscopy and post-embedment immunogold procedures with polyclonal
antibodies have confirmed that CRALBP is found in the cytoplasm of
Muller cells and RPE (figures 12-14). While the cytoplasm is
labeled by the antibody, the nuclei and cytoplasmic organelles do
not show labeling above background. Recent studies of the
localization of CRBP demonstrate that this binding protein is also
present in the cytoplasm of RPE and Muller cells (Bok et al., 1984;

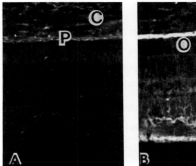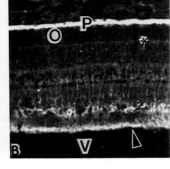

Figure 9. Immunocytochemical localization of CRALBP in bovine retina using a monoclonal antibody. Light micrographs of bovine retina processed by the indirect immunofluorescence technique. A, section treated with control non-specific antibody. The overlying retinal pigment epithelium (P) shows autofluorescence characteristic of lipofuscin. No specific staining is found, although the connective tissue of the choroid (C) is stained above background level. B, section treated with anti-CRALBP monoclonal antibody. Note specific staining of cytoplasm of retinal pigment epithelium (P) and Muller glial cells, whose processes span the retina. O, unstained outer segment layer; *, external limiting membrane; arrow heads, internal limiting membrane; V, vitreous chamber. X 38.

Eisenfeld et al., 1985). Localization of these binding proteins in Muller cells represents a totally unanticipated result and suggests that this glial cell may be the site of one or more of the reactions of the visual cycle.

Early considerations of the role of Muller cells suggest that they play a structural role, providing a scaffold for the orderly array of retinal cells. More recent research indicates that these glial cells are quite active metabolically in spatial buffering of K^+ released from neurons (Newman, 1984) and in neurotransmitter (Sarthy and Lam, 1978) and glycogen (Kuwabara and Cogan, 1966) metabolism. The abundant smooth endoplasmic reticulum in the outer Muller cell processes suggests a role in lipid metabolism, perhaps related to secretion and/or uptake of components of the interphotoreceptor space, facilitated by the numerous Muller microvilli that project into this space (Bunt-Milam and Saari, 1983; Uga and Smelser, 1973). Localization of two retinoid-binding proteins in the Muller cell suggests that this glial cell may play a previously unsuspected role in vitamin A metabolism in the retina.

Independent demonstration of 11-cis-retinaldehyde or retinol in these Muller cells will strengthen the suggestion that they function in metabolism of vitamin A.

In conclusion, study of retinoid-binding proteins has broadened our perspective of reactions of the visual cycle. Water soluble pools of retinoids exist, bound to specific binding proteins. The ability of these to serve as enzyme substrates raises the possibility of their participation in the visual cycle. Metabolic routing of some of the retinoids may be determined by the binding protein with which they are associated. One of the binding proteins, CRALBP, recognizes the 11-cis-configuration with great specificity. In view of the uncertainty about the existence of a stereospecific "retinal isomerase", the ability of this protein to select 11-cis-isomers leads to its consideration as a component of dark regeneration of 11-cis-retinaldehyde. Finally, the presence in Muller cells of two retinoid-binding proteins with their complement of bound retinoids should stimulate further evaluation of the role of these cells in the visual cycle.

ACKNOWLEDGEMENTS

Portions of this work were done in collaboration with D.L. Bredberg, G.G. Garwin and I. Klock. B. Clifton, R. Jones and P. Siedlak provided photographic assistance. The studies were supported in part by USPH Service grants EY-00343, EY-01311, EY-02317, EY-01730 and EY-07013 and by an unrestricted grant from Research to Prevent Blindness, Inc. Dr. Bunt-Milam is a William and Mary Greve International Research Scholar of Research to Prevent Blindness, Inc..

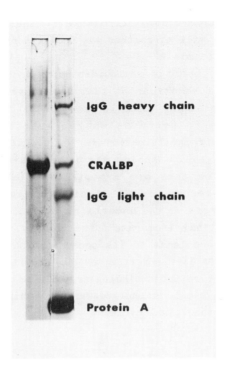

IgG heavy chain

CRALBP

IgG light chain

Protein A

Figure 10. Specificity of monoclonal anti–CRALBP determined by immunopreciptation. A supernatant derived from homogenization of bovine retina was mixed with anti–CRALBP. Antigen–antibody complexes were isolated with protein A–Sepharose and analyzed by SDS–PAGE. Left lane, purified CRALBP; right lane, retinal supernatant and antibody.

Figure 11. A low magnification electron micrograph of the localization of CRALBP in the apical processes of Muller glial cells of the bovine retina. Sections of retina embedded in LR–White resin were treated with anti–CRALBP, followed by a secondary antibody linked covalently to colloidal gold (electron dense dots). Note distribution of gold particles throughout the Muller cell cytoplasm (M) but absence of staining of the photoreceptor inner segments and the interphotoreceptor space (*). Dense granules in the inner segments are primarily ribosomes, which are larger and less electron dense than the 15 nm diameter gold particles. RIS, rod inner segments; R, rod nucleus; CIS, cone inner segment. X 12,500. Bar = 1.0 μm.

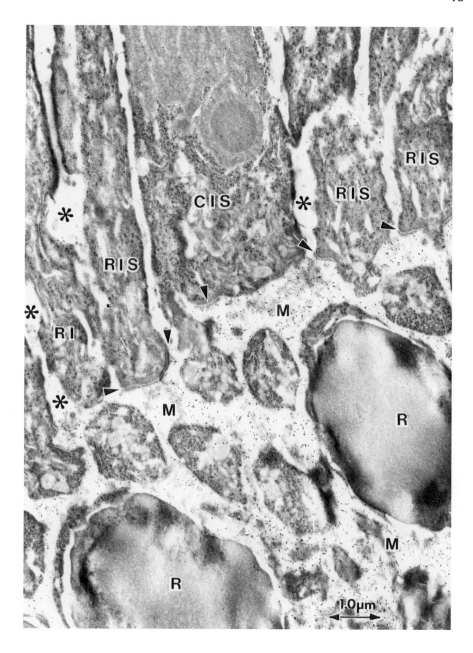

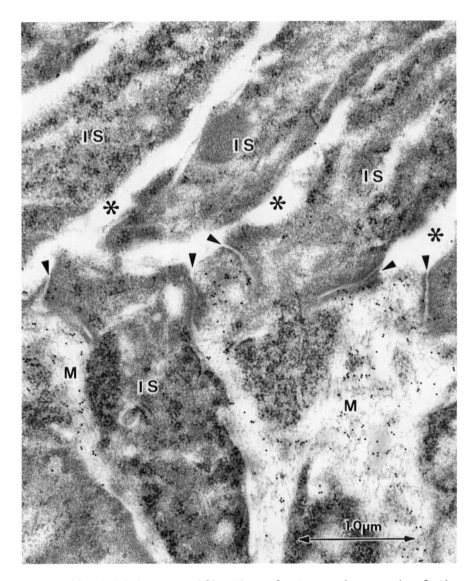

Figure 12. A higher magnification electron micrograph of the localization of CRALBP in the region of the external limiting membrane, a series of intercellular junctions (arrow heads) between the Muller (M) apical processes (labeled specifically by immunogold procedure) and the unlabeled photoreceptor inner segments (IS). *, interphotoreceptor space, which is unlabeled with anti-CRALBP. X 31,500. Bar = 1.0 μm.

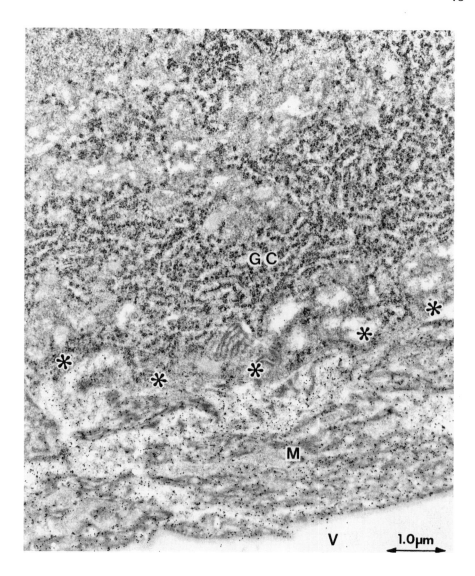

Fig 13. Inner (vitread) surface of the bovine retina stained for CRALBP by the indirect immunogold procedure. Note distribution of gold particles over the cytoplasm of the Muller end feet (M) and absence of label over vitreous humor (V) and a ganglion cell (GC). Since the tissue was not postfixed with osmium tetroxide, surface membranes are not visible. The boundary between the unlabeled ganglion cell cytoplasm (which contains rough endoplasmic reticulum) and the Muller cytoplasm (which contains filaments) is difficult to discern, and is indicated here by *. X15,000. Bar = 1.0 μm.

REFERENCES

Adler, A.J. and K.J. Martin. 1982. Retinol-binding proteins in bovine interphotoreceptor matrix. Biochem. Biophys. Res. Commun. 108: 1601-1608.

Alpern, M. 1971. Rhodopsin kinetics in the human eye. J. Physiol. 217: 447-471.

Andrews, J.S. and S. Futterman. 1964. Metabolism of the retina. V. The role of microsomes in vitamin A esterification in the visual cycle. J. Biol. Chem. 239: 4073-4076.

Berman, E.R., Horowitz, J., Segal, N., Fisher, S. and L. Feeney-Burns. 1980a. Enzymatic esterification of vitamin A in the pigment epithelium of bovine retina. Biochim. Biophys. Acta 630: 35-46.

Berman, E.R., Segal, N., Schneider, A. and L. Feeney. 1980b. An hypothesis for a vitamin A cycle in the pigment epithelium of bovine retina. Neurochem. 1: 113-122.

Bernstein, P.S., Lichtman, J.R. and R.R. Rando. 1985. Nonstereospecific biosynthesis of 11-cis-retinal in the eye. Biochemistry 24: 487-492.

Bok, D. and J. Heller. 1976. Transport of retinol from the blood to the retina: an autoradiographic study of the pigment epithelial cell surface receptor for plasma retinol-binding protein.

Bok, D., Ong, D.E., and R. Chytil. 1984. Immunocytochemical localization of cellular retinol binding protein in the rat retina. Invest. Ophthalmol. Visual Sci. 25: 877-883.

Bridges, C.D.B. 1976. Vitamin A and the role of the pigment epithelium during bleaching and regeneration of rhodopsin in the frog eye. Exp. Eye Res. 22: 435-455.

Bridges, C.D.B. 1984. Retinoids in photosensitive systems. In "The retinoids". (M.B. Sporn, A.B. Roberts and D.S. Goodman, eds.). Vol 2: 125-176.

Bunt-Milam, A.H. and J.C. Saari. 1983. Immunocytochemical localization of two retinoid-binding proteins in vertebrate retina. J. Cell Biol. 97: 703-712.

Bunt-Milam, A.H., Saari, J.C., and D. L. Bredberg. 1985. Characterization of the interstitial space: immunocytochemical and biochemical studies. In "The Interphotoreceptor Matrix in Health and Disease". (C.D. Bridges and A.J. Adler, eds.). Alan R. Liss, Inc. New York. pp. 151-170.

Chader, G.J. 1982. Retinoids in ocular tissues: Binding proteins, transport and mechanism of action. In "Cell Biology of the Eye" (D.McDevitt, ed.). Academic Press, Inc., New York. pp. 377-433.

Chytil, F. and D.E. Ong. 1982. Cellular retinoid binding proteins. In "The retinoids". (M.B. Sporn, A.B. Roberts and D.S. Goodman, eds.). Vol. 2: 89-123.

Crabb, J.W. and J.C. Saari. 1981. N-terminal sequence homology among retinoid-binding proteins from bovine retina. FEBS Letters 130: 15-18.

Dartnall, H.J.A. 1972. Photosensitivity. Handbook of Sensory Physiology Vol. VII/1: 122-145.

Dowling, J.E. 1960. Chemistry of visual adaptation in the rat. Nature 188: 114-118.

Fong, S.-L., Liou, G.I., Landers, R.A., Alvarez, R.A. and C.D. Bridges. 1984. Purification and characterization of a retinol-binding glycoprotein synthesized and secreted by bovine neural retina. J. Biol. Chem. 259: 6534-6542.

Eisenfeld, A.J., Bunt-Milam, A.H. and J.C. Saari. 1985. Localization of retinoid-binding proteins in developing rat retina. Exp. Eye Res. in press.

Futterman, S. 1963. Metabolism of the retina. III. The role of reduced triphosphopyridine nucleotide in the visual cycle. J. Biol. Chem. 238: 1145-1150.

Futterman, S. 1974. Recent studies on a possible mechanism for visual pigment regeneration. Exp. Eye Res. 18: 89-96.

Futterman, S. and J.C. Saari. 1978. Occurrence of 11-cis-retinal-binding proteins restricted to retina. Invest. Ophthalmol. Visual Sci. 16: 768-771.

Futterman, S., Saari, J.C. and S. Blair. 1977. Occurrence of a binding protein for 11-cis-retinal in retina. J. Biol. Chem. 252: 3267-3271.

Hubbard, R. 1956. Retinene isomerase.J. Gen. Physiol. 39: 935-962.

Hubbard, R. and A.D. Colman. 1959. Vitamin A content of the frog eye during light and dark adaptation. Science 130: 977-978.

Julia, P., Farres, J. and X. Pares. 1983. Purification and partial characterization of a rat retina alcohol dehydrogenase active with ethanol and retinol. Biochem. J. 213: 547-550.

Krinsky, J.I. 1958. The enzymatic esterification of vitamin A. J. Biol. Chem. 232: 881-814.

Kuwabara, T. and D. G. Cogan. 1966. Tetrazolium studies on the retina. III. Activity of metabolic intermediates and miscellaneous substrates. J. Histochem. Cytochem. 8: 214-224.

Lai, Y.L., Wiggert, B., Liu, Y.P. and G.J. Chader. 1982. Interphotoreceptor retinol-binding proteins: possible transport vehicles between compartments of the retina. Nature 298: 848-849.

Lehninger, A.L. 1982. Principles of Biochemistry. Worth Publishers, Inc. New York. p. 266.

Lion, F., Rotmans, J.P., Daemen, F.J.M. and S.L. Bonting. 1975. Biochemical aspects of the visual cycle. XXVII. Stereospecificity of the ocular retinol dehydrogenases and the visual cycle. Biochim. Biophys. Acta 384: 383-392.

Liou, G.I., Bridges, C.D.B., Fong, S.-L. Alvarez, R.A. and F. Fernandez-Gonzalez. 1982. Vitamin A transport between retina and pigment epithelium: an interstitial protein carrying endogenous retinol. Vision Res. 22: 1457-1467.

Newman, E.A. 1984. Regional specialization of retinal glial cell membrane. Nature 308: 155-157.

Ostapenko, I.A. and V.V. Furayev, 1973. 9-cis isomerization of all-trans retinal during in vitro regeneration of visual pigment. Nature New Biol. 243: 185-186.

Perlman, J.I., Nodes, B.R. and D.R. Pepperberg. 1982. Utilization of retinoids in the bullfrog retina. J. Gen Physiol. 80: 885-913.

Rando, R.R. and A. Chang. 1983. Studies on the catalyzed interconversions of vitamin A derivatives. J. Am. Chem. Soc. 105: 2879-2882.

Rotmans, J.P., Daemen, F.J.M. and S.L. Bonting. 1972. XIX. Formation of isorhodopsin from photolyzed rhodopsin by bacterial action. Biochim. Biophys. Acta 267: 583-587.

Saari, J.C. and L. Bredberg. 1982. Enzymatic reduction of 11-cis-retinal bound to cellular retinal-binding protein. Biochim. Biophys. Acta 716: 266-272.

Saari, J.C., Bredberg, L. and G.G. Garwin. 1982. Identification of the endogenous retinoids associated with three cellular retinoid-binding proteins from bovine retina and retinal pigment epithelium. J. Biol. Chem. 257: 13329-13333.

Saari, J.C., Bunt-Milam, A.H., Bredberg, D.L. and G.G. Garwin. 1984. Properties and immunocytochemical localization of three retinoid-binding proteins from bovine retina. Vision Res. 24: 1595-1603.

Sarthy, P.V. and A.H. Bunt. 1982. The ultrastructure of isolated glial (Muller) cells from the turtle retina. Anat. Rec. 202: 275-283.

Sarthy, P.V. and D.M.K. Lam. 1978. Biochemical studies of isolated glial (Muller) cells from the turtle retina. J. Cell Biol. 78: 675-684.

Stubbs, G.W., Saari, J.C. and S. Futterman. 1979. 11-cis-retinal-binding protein from bovine retina. J. Biol. Chem. 254: 8529-8533.

Sundelin, J., Das, S.R., Eriksson, U., Rask, L. and P.A. Peterson. 1985. The primary structure of bovine cellular retinoic acid-binding protein. J. Biol. Chem. 260: 6494-6499.

Uga, S. and G.K. Smelser. 1973. Comparative study of the fine structure of retinal Muller cells in various vertebrates. Invest. Ophthalmol. 12: 434-448.

Vance, D.E. 1983. Metabolism of steroids and lipoproteins. In "Biochemistry" (G. Zubay, coordinating author). Addison-Wesley Pub. Co., Reading, MA. p. 566.

Wald, G. 1968. The molecular basis of visual excitation. Science 162: 230-239.

Wald, G. and R. Hubbard. 1949. The reduction of retinene to vitamin A in vitro. J. Gen. Physiol. 32: 367-389.

Wiggert, B., Bergsma, D.R., Lewis, M. and G.J. Chader. 1977. Vitamin A receptors: retinol binding in neural retina and pigment epithelium. J. Neurochem. 29: 947-954.

Wiggert, B.O. and G.J. Chader. 1975. A receptor for retinol in developing retina and pigment epithelium. Exp. Eye Res. 21: 143-151.

Yoshikami, S. and G.N. Noll. 1978. Isolated retinas synthesize visual pigments from retinol congeners delivered by liposomes. Science 200: 1393-1395.

Zimmerman, W.F., Lion, F., Daemen, F.J.M. and S.L. Bonting. 1975. Biochemical aspects of the visual process. XXX. Distribution of stereospecific retinol dehydrogenase activities in subcellular fractions of bovine retina and pigment epithelium. Exp. Eye Res. 21: 353-332.

Roles of Intra- and Extracellular Retinoid-Binding Proteins in the Visual Cycle

C. D. B. Bridges

All light-initiated events in the eye depend on the absorption of quanta of incident radiation by visual pigment molecules in the photoreceptor outer segments. Visual pigments are transmembrane glycoproteins that have retinaldehyde (vitamin A_1 aldehyde) or 3,4-didehydroretinaldehyde (vitamin A_2 aldehyde) as their prosthetic groups. In the rhodopsins, the prosthetic group is 11-cis retinaldehyde, and in the porphyropsins, which are present in the eyes of amphibians and fishes, this is replaced by 11-cis-3,4-didehydroretinaldehyde (for review, see Bridges, 1972).

In bovine rhodopsin, 11-cis retinaldehyde is covalently attached to Lys-296 of the glycoprotein moiety (opsin) by an aldimine linkage (Wang et al., 1980; for review, see Hargrave et al., 1983). Light isomerizes the 11-cis isomer to all-trans, and sets in train a succession of intermediate products, many of which have only a transient existence at room temperature (Wald, 1968). The structures of these intermediates are still not understood. They include hypsorhodopsin, bathorhodopsin, lumirhodopsin, metarhodopsins I and II, pararhodopsin and N-retinylideneopsin (Yoshizawa and Shichida, 1982; Applebury & Rentzepis, 1982). The final intermediate, usually identified as N-retinylideneopsin, is unstable at physiological pH and temperature, and hydrolyzes to opsin and all-trans retinaldehyde. In the isolated human retina at 36°C, free retinaldehyde has a half-life of 23 seconds (Baumann & Bender, 1973), and is reduced to all-trans retinol. The oxidoreductase involved in this reaction differs from the zinc-dependent liver alcohol dehydrogenase (which can also reduce retinaldehyde to retinol) in that it is membrane-bound, requires NADP rather than NAD as a cofactor, and is more specific in that it has preference for the all–trans isomer and cannot use ethanol as a substrate (Futterman, 1963; Lion et al., 1975; Bridges, 1977).

Because it is the precursor of the retinaldehyde prosthetic group of rhodopsin, retinol is need to maintain the visual process. Retinoids must therefore be supplied to the eye, stored in its tissues, and transported within and between its cells.

DELIVERY OF VITAMIN A TO THE EYE: SERUM RETINOL–BINDING PROTEIN (RBP)

In the bloodstream, all–trans retinol is carried by a 21,000 dalton protein known as serum retinol-binding protein (for review, see Goodman, 1984). This protein circulates as a 1:1 complex with transthyretin (prealbumin). RBP solubilizes retinol, protects it from oxidative degradation and provides the means for targeting its delivery to cells that have RBP receptors on their plasma membranes. Receptors for RBP have been found on the surfaces of intestinal mucosa, testicular and corneal epithelial cells, as well as on the basement membranes of the pigment epithelium (RPE) cells (Rask & Peterson, 1976; Bhat & Cama, 1979; Bok & Heller, 1976: McGuire et al., 1979; Rask et al., 1980).

STORAGE OF VITAMIN A IN THE EYE

The highest concentrations of vitamin A occur in the liver and RPE. The RPE from a pair of human eyes contains 4.5 ± 2.6 µg of vitamin A. This is between 2 and 3 molar equivalents of rhodopsin in the retina (Bridges et al., 1982).

Most of the vitamin A in the RPE consists of retinyl palmitate and stearate in the approximate ratio of 5:1. As much as 75% of these esters may be in the 11–cis configuration. A typical chromatogram obtained by high–performance liquid chromatography of the retinyl esters stored in human RPE is illustrated in Fig. 1. Two major peaks are evident, representing 11–cis retinyl palmitate (peak 2) and all–trans retinyl palmitate (peak 4). The shoulders on the leading edges of these peaks (1,3) are attributable to the corresponding stearates (these were not fully resolved on the normal–phase column which was best suited for isomer separation).

In the dark–adapted eyes of most animals, the amount of vitamin A in the RPE represents between 1 and 6 mol equivalents of the visual pigment in the retina. It is not known whether the magnitude of these stores depends on vitamin A nutriture, but it is probable

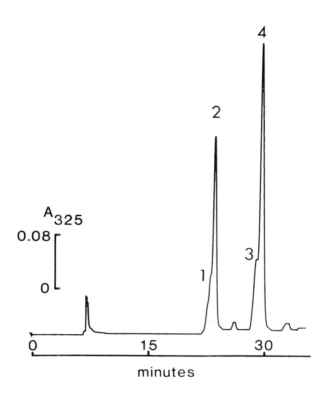

Figure 1. Vitamin A esters in the human pigment epithelium: high-performance liquid chromatography on normal-phase columns.
11-cis retinyl palmitate, peak 2; all-trans retinyl palmitate, peak 4; the shoulders labeled 1 and 3 indicate the positions of the 11-cis and all-trans retinyl stearates, respectively. Columns, mobile phase, equipment and other conditions are as described by Bridges et al. (1982).

that they serve as a reserve that protects the visual system from depletion under conditions of dietary deficiency.

EXTRACELLULAR TRANSPORT OF VITAMIN A BETWEEN PIGMENT EPITHELIUM AND RETINA: INTERSTITIAL RETINOL-BINDING PROTEIN (IRBP)

The pigment epithelium is not only a source of vitamin A for the retina. The all-trans retinol that is generated when rhodopsin is bleached flows outward from the rod outer segments (ROS) and enters the RPE cells through their apical membranes. It accumulates in the form of all-trans retinyl esters (Dowling, 1960; Bridges,

1976b). The transfer of retinol between the ROS and the RPE (Fig. 2) entails passage through the interphotoreceptor matrix, which contains interstitial retinol-binding protein (IRBP; Liou et al., 1982a,b), a glycoprotein that in cattle has been found to carry endogenous all-trans and 11-cis retinol (Adler & Martin, 1982; Liou et al,. 1982a,b). IRBP is a major component of the interphotoreceptor matrix; in cattle it is present at a concentration of 30-100 µM (Fong et al., 1984b).

It has been suggested that IRBP is the transport vehicle that shuttles retinol between the cells of the retina and the RPE during the visual cycle (Fig. 2 ; Adler & Martin, 1982; Liou et al., 1982a,b; Lai et al., 1982; Pfeffer et al., 1983; Fong et al., 1984a,b; Bridges, 1984; Bridges et al., 1984; Saari et al., 1984). It is therefore significant that its first appearance during development has been noted when the inner segments are being formed, but prior to the time when rhodopsin and 11-cis retinyl esters are first detectable in the eye (Carter-Dawson et al., 1986).

IRBP has been purified and characterized from human, bovine, monkey and rat eyes (Fong et al., 1984a,b; Alder et al., 1985; Gonzalez-Fernandez et al., 1985; Redmond et al., 1985; Saari et al., 1985). Bovine IRBP has an apparent molecular weight of 144,000 daltons on sodium dodecyl sulfate polyacrylamide gels (Fig. 3), binds about 2 molecules of all-trans retinol, and has 4-5 carbohydrate chains that appear to consist of fucosylated hybrid-and complex-type oligosaccharides (Fong et al., 1984a, 1985a; Taniguchi et al., 1986). It exhibits an anomalously high molecular weight of 250,000 daltons on gel-filtration columns because it is an elongated molecule with an axial ratio of about 8:1 (Alder et al., 1985; Saari et al., 1985). Human IRBP has a lower apparent molecular weight on SDS gels (Fig. 3) and gel filtration columns. Its amino acid composition and complement of oligosaccharide chains are similar to bovine IRBP (Fong et al., 1984a), and it is immunologically reactive with anti-bovine IRBP antibodies.

As might be expected from its suggested role in the visual cycle, IRBP is universally distributed in the IPM of the major vertebrate classes, which include the Mammalia, Aves, Reptilia, Osteichthyes, Chondrichthyes and Amphibia (Bridges et al., 1984; Fong et al., 1985b; Bridges et al., 1986c). This was demonstrated by electrophoretically transferring the IPM proteins from sodium dodecyl sulfate polyacrylamide gels to nitrocellulose sheets, then

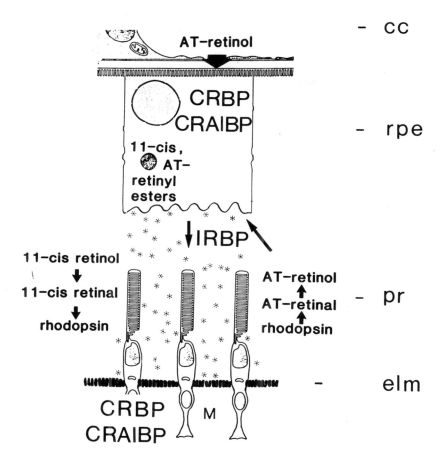

Figure 2. Retinoid-binding proteins and the visual cycle. Endogenous ligands shown below in parentheses.
RBP, plasma retinol-binding proteins (all-trans retinol); CRBP, cellular retinol-binding protein (all-trans retinol); IRBP, interstitial retinol-binding protein (all-trans and 11-cis retinol; traces of isomers of retinaldahyde); CRAlBP, cellular retinaldehyde-binding protein (11-cis retinol and 11-cis retinaldehyde). CRABP, cellular retinoic acid-binding protein, is not shown. Retinoic acid is not convertible to retinol or retinaldehyde and hence cannot participate directly in the visual cycle. cc, choriocapillaris; rpe, retinal pigment epithelium; pr, photoreceptors; elm, external limiting membrane; M, Muller cell. The interphotoreceptor matrix is indicated with asterisks.

incubating the sheets with rabbit antibovine IRBP immunoglobulins followed by horseradish peroxidase-conjugated goat anti-rabbit IgG. In most of these animals, IRBP-like immunoreactivity was observed in

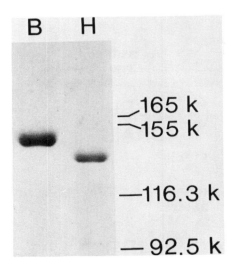

Figure 3. Human and bovine IRBP: comparison of electrophoretic mobilities of the purified proteins by sodium dodecyl sulfate polyacrylamide gel electrophoresis. B, bovine IRBP (molecular weight = 144k); H, human IRBP (molecular weight = 135k). Position of molecular weight markers shown on right. Conditions as described by Fong et al. (1984a,b).

proteins with molecular weight of $134,200\pm8,600$ (n=17). In the Osteichthyes, however, the molecular weight was about half of this value ($67,600\pm2,700$, n=8), leading to the suggestion that there may have been at least one gene duplication in the evolution of this protein. Such an event, perhaps involving the gene coding for a primordial retinoid-binding protein, might explain the ability of IRBP to bind two molecules of retinol (Fong et al., 1984b; Saari et al., 1985), possibly in two discrete binding sites.

It has been shown that IRBP is synthesized by the retina, not by the RPE. This has been demonstrated in experiments where isolated retinas from cattle, rats, humans, monkeys, ranid frogs and Xenopus have been incubated with labeled sugar (fucose, glucosamine) or amino acid precursors (Bridges et al., 1986c; Fong et al., 1984a,b; Wiggert et al., 1984; Rayborn et al., 1984; Gonzalez-Fernandez et al., 1984,1985; Hollyfield et al., 1985). In each case, labeled protein with an electrophoretic mobility identical with that of IRBP and immunoprecipitable with rabbit antibovine IRBP immunoglobulin was secreted into the incubation medium. Secretion was not prevented by tunicamycin, an antibiotic that inhibits the assembly of oligosaccharides linked N-glycosidically to proteins (Fong et al., 1984a). The medium in which pigment epithelium cells had been incubated contained no trace of labeled IRBP.

Of the retinal cells that border the subretinal space, the photoreceptors rather than the Muller (glial) cells appear to be the source of IRBP. Thus, these cells appear to be involved in the biosynthesis of two major glycoproteins that have an important role in the biochemistry of the visual process, viz. rhodopsin and IRBP.

The evidence that the photoreceptors produce IRBP has been presented by Bridges (1985), and is summarized below.

(1) Gonzalez-Fernandez et al. (1984) showed that when RCS rat retinas lose their photoreceptors, but are otherwise normal in their morphology, they also lose their ability to synthesize IRBP.

(2) Oka et al. (1985) showed that cultures of adult human retinal cells secreted IRBP into the medium only when neuronal cells (some of which were presumably photo-receptors) were present. Cultures containing retinal glial cells alone did not.

(3) Hollyfield et al. (1985) used pulse-chase autoradiography of isolated human retinas incubated with (^{3}H)-fucose to demonstrate that a fucosylated compound was synthesized and secreted by the photoreceptors (cf. Feeney, 1973). In these experiments, (^{3}H)-IRBP was subsequently recovered from the medium.

The search for IRBP mRNA within the photoreceptor inner segments is now being carried out with cDNA probes derived from the coding region of the IRBP gene and should provide an unequivocal answer to this question (Liou et al., in preparation). Although in the developing normal retina IRBP is expressed when the inner segments start to differentiate (Carter-Dawson et al., 1986), it should be noted that large amounts of IRBP have been detected in undifferentiated retinoblastoma tumors. It has also been shown that the cells from these tumors are capable of synthesizing IRBP from radiolabeled precursors in vitro and secreting it. These finding suggest that regulation of the IRBP gene in this neuroectodermal neoplasm may be defective (Bridges et al., 1985).

INTRACELLULAR TRANSPORT OF VITAMIN A - CELLULAR RETINOID-BINDING PROTEINS (FIG.2)

When retinol passes from the plasma membrane to its intracellular sites of esterification and utilization, it apparently becomes bound to an intracellular protein, cellular retinol-binding protein (CRBP; Ong & Chytil, 1978). CRBP is distinct from RBP immunologically, spectroscopically, by its lower molecular weight (15,000 daltons) and by its failure to complex with transthyretin. CRBP occurs in many tissues (for review, see Chytil & Ong, 1984), but most of the CRBP in the eye is found in the RPE. A small quantity is associated with the cells of the neural retina (Bok et al., 1985; Saari et al., 1978). Its endogenous ligand is exclusively all-<u>trans</u> retinol (Saari et al., 1982).

Binding proteins for other retinoids are also present in the retina and RPE: they include cellular retinoic acid binding protein (CRABP; Saari et al., 1978) and cellular retinal-binding protein (CRAlBP; Stubbs et al., 1979). CRABP occurs in many tissues as well as the retina (Chytil & Ong, 1984). It is absent from the RPE. Its endogenous ligand consists of all-<u>trans</u> retinoic acid. CRAlBP is restricted to the RPE and the Muller cells of the retina (Bunt-Milam & Saari, 1983; this volume). Its endogenous ligands consist of 11-<u>cis</u> retinaldehyde and 11-<u>cis</u> retinol (Saari et al., 1982).

REGENERATION OF RHODOPSIN-RETINOID-BINDING PROTEINS AND THE DISTRIBUTION OF VITAMIN A ISOMERS

Rats and frogs have been the main experimental animals for studies on the visual cycle (Dowling, 1960; Zimmerman, 1974; Bridges, 1976b). Dark-adapted rats have very little vitamin A in the RPE, but frogs store about 2 moles of retinyl palmitate per mole of rhodopsin in the retina. After 24 hours in darkness about half of this ester may be 11-<u>cis</u> (Bridges, 1976b). In both species it has been demonstrated that when a large fraction of rhodopsin is bleached by intense light-adaptation, there is corresponding accumulation of all-<u>trans</u> retinyl ester in the RPE. In frogs, it has also been shown that pre-existing supplies of 11-<u>cis</u> retinyl ester are consumed over several cycles of bleaching and regeneration (Bridges, 1976b). Therefore, at the end of a period of strong light-adaptation, rats and frogs are faced with the task of carrying out an overall conversion of all-<u>trans</u> retinyl ester to 11-<u>cis</u> retinaldehyde. The retinoid that is isomerized, and where this isomerization occurs, has not been determined. Since photo-isomerization of all-<u>trans</u> retinoid does not appear to play a role in the physiological process of dark-adaptation in vertebrates, a dark reaction must be sought (see Bridges, 1984, for review). Furthermore, recent observations (Bunt-Milam & Saari, 1983) have suggested that the apparently straightforward exchange of retinoids between the ROS and RPE may be more complicated, and perhaps involves the Muller cells.

The following observations provide an important insight into how the visual cycle functions.

(1) CRAlBP, which binds 11-<u>cis</u> retinoids, occurs only in the RPE and Muller cells (Bunt-Milam & Saari, 1983).

(2) The major endogenous ligand of CRAlBP is 11-<u>cis</u> retinaldehyde in the RPE, but mainly 11-<u>cis</u> retinol in the retina (Saari et al., 1982; see also Liou et al., 1982b; Bridges et al., 1984): membrane fractions from the retina (but not the RPE) also contain 11-<u>cis</u> retinol (Bridges, 1976a; Bridges et al., 1984).

(3) Eleven-<u>cis</u> retinol oxidoreductase, which is necessary to form 11-<u>cis</u> retinol, is absent from the retina but present in the RPE (Zimmerman et al. 1975; Lion et al., 1975).

The RPE is generally acknowledged to have a role in visual pigment regeneration, yet no isomerase has been demonstrated in fresh or cultured RPE cells (Fong et al., 1983; Flood et al., 1983). Therefore, it has been suggested that the site of isomerization must lie in the retina (Bridges, 1976a,b; Fong et al., 1983; Flood et al., 1983). The difficulty with this idea is that, aside from one unconfirmed observation in the rat (Cone & Brown, 1969), isolated retinas do not regenerate rhodopsin.

The need for the RPE in rhodopsin regeneration, even though the site of isomerization is in the retina, has been explained in the following way (Bridges et al., 1984). The 11-<u>cis</u> retinol found in the retina may have been formed in the Muller cells, perhaps after IRBP has transported all-<u>trans</u> retinol from the ROS to their apical membranes. Yet, as noted above, 11-<u>cis</u> retinaldehyde cannot be formed from 11-<u>cis</u> retinol in the retina because this tissue lacks 11-<u>cis</u> retinol oxidoreductase (Pepperberg & Masland, 1978; Yoshikami & Noll, 1978; Zimmerman et al., 1975; Lion et al., 1975). The occurrence of this stereospecific enzyme in the RPE (Zimmerman et al.,1975; Lion et al., 1975) coupled with the predominance of 11-cis retinaldehyde bound to CRAlBP in the RPE cytosol, suggests that the role of the RPE in the visual cycle is to convert 11-cis retinol to 11-cis retinaldehyde. As summarized in Fig. 4, this hypothesis requires that 11-cis retinol must be delivered to the RPE and that 11-cis retinaldehyde is returned to the ROS. In the interphotoreceptor matrix IRBP is probably implicated in the transport of both these 11-cis retinoids as well as all-trans retinol (Fig. 5).

In the scheme in Fig. 4, any 11-cis retinol that is not immediately utilized by the pathway leading to rhodopsin regeneration would be esterified and stored, so accounting for the observed accumulation of 11-cis retinyl esters in dark-adapted eyes (Krinsky, 1958; Alvarez et al., 1981). This pool may also exchange with opsin molecules in the dark (Bridges & Yoshikami, 1970; Defoe & Bok, 1983). Large quantities of all-trans retinol released by strong illumination would bind to IRBP (Fig. 5) at the surface of the ROS, and could by-pass the isomerization reaction by traveling directly to the RPE.

In summary, 11-cis retinol in the mammalian retina may be generated from all-trans retinol in the Muller cells, but it must first be transported to the RPE for conversion to 11-cis retinaldehyde before it is returned to the ROS. IRBP is probably implicated in transporting retinoids through the interphotoreceptor matrix between the ROS, Muller cells and RPE.

ACKNOWLEDGEMENTS

Work reported in this paper was supported by the Retina Research Foundation of Houston, the National Eye Institute and Research to Prevent Blindness, Inc.

NOTE ADDED IN PROOF

Recent work in the author's laboratory has identified in frog, rats and cattle an eye-specific membrane bound enzyme that isomerizes all-trans retinol to 11-cis retinol. It is localized in the pigment epithelium. Reaction 2 in Fig. 4 therefore occurs in the pigment epithelium, not the Mueller cells. This new finding accounts for the role of the pigment epithelium in rhodopsin regeneration but raises an important question concerning the role of the Mueller cells in the visual cycle.

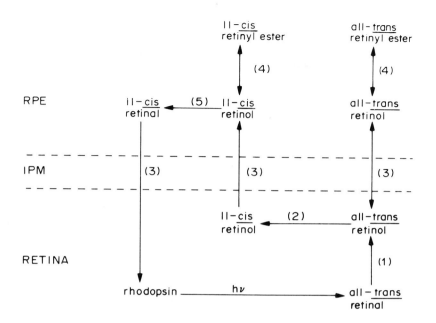

Figure 4.   Proposed visual cycle in the mammalian eye (after Bridges et al., 1984).   The numbered steps are as follows.   (1) all-trans retinol oxidoreductase (Futterman, 1963; Lion et al., 1975; Zimmerman et al., 1975).   (2) presumptive isomerase.   (3) transport of retinoid through the IPM, probably bound to IRBP.   (4) esterifying enzyme which may be a reversed hydrolase rather than an acyl-CoA retinol acyl transferase (Krinsky, 1958; Flood et al., 1983; Fong et al., 1983; Bridges et al., 1986b).   (5) 11-cis retinol oxidoreductase (Lion et al., 1975; Zimmerman et al., 1975).

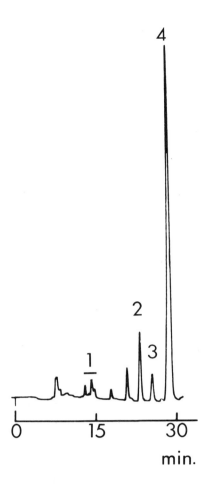

Figure 5.   Endogenous retinoids extracted from bovine IRBP: high-performance liquid chromatography.   11-<u>cis</u> retinol, peak 2; 13-<u>cis</u> retinol, peak 3; all-<u>trans</u> retinol, peak 4.   The group of peaks labeled 1 represents small quantities of isomers of retinaldehyde including the 11-<u>cis</u> and all-<u>trans</u>.   Unnumbered peaks were not identified.   Columns, mobile phase, equipment and other conditions have been described by Bridges et al. (1984).

REFERENCES

Adler, A.J. and K.J. Martin. 1982. Retinol-binding proteins in bovine interphotoreceptor matrix. Biochem. Biophys. Res. Commun. 108:1601-1608.

Adler, A.J., C.D. Evans and W.F. Stafford III. 1985. Molecular properties of bovine interphotoreceptor retinol-binding protein. J. Biol. Chem. 260:4850-4855.

Alvarez, R.A., C.D.B. Bridges and S.-F. Fong. 1981. High-pressure liquid chromatography of fatty acid esters of retinol isomers-analysis of retinyl esters stored in the eye. Invest. Ophthalmol. Vis. Sci. 20:304-313.

Applebury, M.L. and P. M. Rentzepis. 1982. Picosecond spectroscopy of visual pigments. Methods in Enzymology 81:354-368.

Baumann, C. and S. Bender. 1973. Kinetics of rhodopsin bleaching in the isolated retina. J. Physiol. 235:761-773.

Bhat, M.K. and H.R. Cama. 1979. Gonadal cell surface receptor for plasma retinol binding protein. A method for its radioassay and studies on its level during spermatogenesis. Biochim. Biophys. Acta 578:273-281.

Bok, D. and J. Heller. 1976. Transport of retinol from the blood to the retina: an autoradiographic study of the pigment epithelial cell surface receptor for plasma retinol binding protein. Exp. Eye Res. 22:395-402.

Bok, D., D.E. Ong and F. Chytil. 1984. Immunocytochemical localization of cellular retinol binding protein in the rat retina. Invest. Ophthal. & Vis. Sci. 25:877-883.

Bridges, C.D.B. 1972. "The rhodopsin-porphyropsin visual system." In Handbook of Sensory Physiology 7/1A. Ed. by H.J.A. Dartnall. Springer-Verlag, Berlin, pp. 418-480.

Bridges, C.D.B. 1976a. 11-cis vitamin A in dark-adapted rod outer segments is a probable source of prosthetic groups for rhodopsin biosynthesis. Nature 259:247-248.

Bridges, C.D.B. 1976b. Vitamin A and the role of the pigment epithelium during bleaching and regeneration of rhodopsin in the frog eye. Exp. Eye Res. 22:435-455.

Bridges, C.D.B. 1977. Rhodopsin regeneration in rod outer segments: Utilization of 11-cis retinal and retinol. Exp. Eye Res. 24:571-580.

Bridges, C.D.B. 1984. "Retinoids in photosensitive systems." In The Retinoids. Ed. by M.B. Sporn, A.B. Roberts and DeWitt S. Goodman. Academic Press, New York, pp. 125-176.

Bridges, C.D.B. 1985. "The interphotoreceptor matrix-functions and possible role in hereditary retinal degenerations." In The Interphotoreceptor Matrix in Health and Disease. Ed. by C. David Bridges and Alice J. Adler. Alan R. Liss, New York. pp. 195-212.

Bridges, C.D.B. and S. Yoshikami. 1970. Uptake of tritiated retinaldehyde by the visual pigment of dark-adapted rats. Nature 221:275-276.

Bridges, C.D.B., R.A. Alvarez and S.-L. Fong. 1982. Vitamin A in human eyes- amount, distribution and composition. Invest. Ophthal. & Vis. Sci. 22:706-714.

Bridges, C.D.B., S.-L. Fong, G.I. Liou, R.A. Landers, F. Gonzalez-Fernandez, P. Glazebrook, and D.M.K. Lam. 1983. A retinol-binding glycoprotein synthesized and secreted by mammalian neural retina. Soc. Neurosci. Abstr. 9:345.

Bridges, C.D.B., R.A. Alvarez, S.-L. Fong, F. Gonzalez-Fernandez, D.M.K. Lam and G.I. Liou. 1984. Visual cycle in the mammalian eye: retinoid-binding proteins and the distribution of 11-cis retinoids. Vision Res. 24:1581-1594.

Bridges, C.D.B., S.-L. Fong, R.A. Landers, G.I. Liou and R.L. Font. 1985. Interstitial retinol-binding protein (IRBP) in retinoblastoma. Neurochem. Int. 7:875-881.

Bridges, C.D.B., T. Peters, J.E. Smith, D.S. Goodman, S.-L. Fong, M.D. Griswold and N.A. Musto. 1986a. Biosynthesis and secretion of transport proteins-interstitial, and serum retinol-binding proteins, transthyretin, transferrin, serum albumin and extracellular sex steroid binding proteins. Fed. Proc. 45:2291-2303.

Bridges, C.D.B., M.S. Oka, S.-L. Fong, G.I. Liou and R.A. Alvarez. 1986b. Retinoid-binding proteins and retinol esterification in cultured pigment epithelium cells. Neurochem. Int. 8:527-534.

Bridges, C.D.B., Liou, G.I., Alvarez, R.a., Landers, R.A., Landry Jr., A.M., and Fong, S.-L. Distribution of interstitial retinol-binding protein (IRBP) in the vertebrates. 1986c. J. Exp. Zool., in press.

Bunt-Milam, A.H. and J.C. Saari. 1983. Immunocytochemical localization of two retinoid-binding proteins in vertebrate retina. J. Cell Biol. 97:703-712.

Carter-Dawson, L., R.A. Alvarez, S.-L. Fong, G.I. Liou, H.G. Sperling and C.D.B. Bridges. 1986. Rhodopsin, 11-cis vitamin A and interstitial retinol-binding protein (IRBP) during retinal development in normal and rd mutant mice. Develop. Biol., in press.

Chytil, F. and D.E. Ong. 1984. "Cellular retinoid-binding proteins." In The Retinoids. Ed. by M.B. Sporn, A.B. Roberts and DeWitt S. Goodman. Academic Press, New York. Vol. 2, pp.89-123.

Cone, R.A. and P.K. Brown. 1969. Spontaneous regeneration of rhodopsin in the isolated rat retina. Nature 221:818-820.

Defoe, D.M. and D. Bok. 1983. Rhodopsin chromophore exchanges among opsin molecules in the dark. Invest. Ophthal. & Vis. Sci. 24:1211-1226.

Dowling, J.E. 1960. Chemistry of visual adaptation in the rat. Nature 188:114-118.

Feeney, L. 1973. Synthesis of interphotoreceptor matrix. I. Autoradiography of (^{3}H-fucose) incorporation. Invest. Ophthal. & Vis. Sci. 12:739-751.

Flood, M.T., C.D.B. Bridges, R.A. Alvarez, W.S. Blaner and P. Gouras. 1983. Vitamin A utilization in human retinal pigment epithelial cells in vitro. Invest. Ophthal. & Vis. Sci. 24:1227-1235.

Fong, S.-L., C.D.B. Bridges and R.A. Alvarez. 1983. Utilization of exogenous retinol by frog pigment epithelium. Vision Res. 23:47-52.

Fong, S.-L., T. Irimura, R.A. Landers and C.D.B. Bridges. 1985a. "The carbohydrate of bovine interstitial retinol-binding protein." In The Interphotoreceptor Matrix in Health and Disease. Ed. by C. David Bridges and Alice J. Adler. Alan R. Liss, New York. pp. 111-128.

Fong, S.-L., K.A. Johnson, G.I. Liou, R.A. Landers, A.M. Landry and C.D.B. Bridges. 1985b. Distribution of interstitial retinol-binding protein in the vertebrates - evidence for gene duplication? Invest. Ophthal. & Vis. Sci. ARVO Supplement 26:17.

Futterman, S. 1963. Metabolism of the retina. III. Role of reduced triphosphopyridine nucleotide in the visual cycle. J. Biol. Chem. 238:1145-1150.

Gonzalez-Fernandez, F., R.A. Landers, P.A. Glazebrook, S.-L. Fong, G.I. Liou, D.M.K. Lam and C.D.B. Bridges. 1984. An extracellular retinol-binding glycoprotein in the eyes of mutant rats with retinal dystrophy - development, localization and biosynthesis. J. Cell Biol. 99:2092-2098.

Gonzalez-Fernandez, F., R.A. Landers, P.A. Glazebrook, S.-L. Fong, G.I. Liou, D.M.K. Lam and C.D.B. Bridges. 1985. An extracellular retinol-binding glycoprotein in the rat eye-characterization, localization and biosynthesis. Neurochem. Int. 7:533-540.

Goodman, D.S. 1984. Plasma retinol-binding protein. In The Retinoids. Ed. by M.B. Sporm, A.B. Roberts and DeWitt S. Goodman. Academic Press, New York. Vol. 2, pp. 41-88.

Hargrave, P.A., J.H. McDowell, D.R. Curtis, J.K. Wang, E. Juszczak, S.-L. Fong, J.K. Mohana Rao and P. Argos. 1983. The structure of bovine rhodopsin. Biophys. Struct. Mech. 9:235-244.

Hollyfield, J.G., S.J. Fliesler, M.E. Rayborn, S.-L. Fong, R.A. Landers and C.D.B. Bridges. 1985. Synthesis and secretion of interstitial retinol-binding protein by the human retina. Invest. Ophthal. & Vis. Sci. 26:58-67.

Krinsky, N.I. 1958. The enzymatic esterification of vitamin A. J. Biol. Chem. 232:881-894.

Lai, Y.L., B. Wiggert, Y.P. Liu and G.J. Chader. 1982. Interphotoreceptor retinol-binding proteins: possible transport vehicles between compartments of the retina. Nature 298:848-849.

Lion, F., J.P. Rotmans, F.J.M. Daemen and S.J. Bonting. 1975. Stereospecificity of ocullar retinol dehydrogenases and the visual cycle. Biochem. Biophys. Acta 384:283-292.

Liou, G.I., C.D.B. Bridges and S.-L. Fong. 1982b. Vitamin A transport between retina and pigment epithelium-an interphoto-receptor matrix protein carrying endogenous retinol (IRBP). Invest. Ophthal. & Vis. Sci. 22:65.

Liou, G.I., C.D.B. Bridges, S.-L. Fong, R.A. Alvarez and F. Gonzalez-Fernandez. 1982b. Vitamin A transport between retina and pigment epithelium - an interstitial protein carrying endogenous retinol (interstitial retinol-binding protein). Vision Res. 22:1457-1468.

McGuire, B.W., M.C. Orgebin-Crist and F. Chytil. 1981. Autoradio-graphic localization of serum retinol-binding protein in rat testis. Endocrinology 108:658-667.

Oka, M.S., J.M. Frederick, R.A. Landers and C.D.B. Bridges. 1985. Adult human retinal cells in culture: identification of cell types and expression of differentiated properties. Exp. Cell Res. 159:127-140.

Ong, D.E. and F. Chytil. 1978. Cellular retinol-binding protein from rat liver. Purification and characterization. J. Biol. Chem. 253:828-832.

Pepperberg, D.R. and R.H. Masland. 1978. Retinal-induced sensiti-zation of light-adapted rabbit photoreceptors. Brain Res. 151:194-200.

Pfeffer, B., B. Wiggert, L. Lee, B. Zonnenberg, D. Newsome, and G. Chader. 1983. The presence of a soluble interphotoreceptor retinol-binding protein (IRBP) in the retinal interphoto-receptor matrix. J. Cell Physiol. 117:333-341.

Rask, L. and P. Peterson. 1976. In vitro uptake of vitamin A from the retinol-binding plasma protein to mucosal epithelial cells from the monkey's small intestine. J. Biol. Chem. 251:6360-6366.

Rask, L., C. Geijer, A. Bill, and P.A. Peterson. 1980. Vitamin A supply of the cornea. Exp. Eye Res. 31:201-211.

Rayborn, M.E., C.D.B. Bridges, R.A. Landers and J.G. Hollyfield. 1984. Synthesis, secretion, localization and immunoreactivity of interstitial retinol-binding protein in Xenopus laevis. Invest Ophthal. & Vis. Sci. ARVO Supplement 25:175.

Redmond, T.M., B. Wiggert, F.A. Robey, N.Y. Nguyen, M.S. Lewis, L. Lee and G.J. Chader. 1985. Isolation and characterization of monkey interphotoreceptor retinoid-binding protein, a unique extracellular matrix component of the retina. Biochemistry 24:787-793.

Saari, J.C., S. Futterman and L. Bredberg. 1978. Cellular retinol- and retinoic-acid binding proteins of bovine retina. Purification and properties. J. Biol. Chem. 253:6432-6436.

Saari, J.C., A.H. Bunt-Milam, D.L. Bredberg and G.G. Garwin. 1984. Properties and immunocytochemical localization of three retinoid-binding proteins from bovine retina. Vision Res. 24:2595-2603.

Saari, J.C., D.C. Teller, J.W. Crabb and L. Bredberg. 1985. Properties of an interphotoreceptor retinoid-binding protein from bovine retina. J. Biol. Chem. 260:195-201.

Stubbs, G.W., J.C. Saari and S. Futterman. 1979. 11-cis-retinal-binding protein from rat retina. Isolation and partial characterization. J. Biol. Chem. 254:8529-8533.

Taniguchi, T., A.J. Adler, T. Mizuochi, N. Kochibe and A. Kobata. 1986. The structures of the asparagine-linked sugar chains of bovine interphotoreceptor retinol-binding protein. Occurrence of fucosylated hybrid-type oligosaccharides. J. Biol. Chem. 261:1730-1736.

Ward, G. 1968. The molecular basis of visual excitation. Nature 219:800-807.

Wang, J.K., J.H. McDowell and P.A. Hargrave. 1980. Site of attachment of 11-cis retinal in bovine rhodopsin. Biochemistry 19:5111-5117.

Wiggert, B., L. Lee, P.J. O'Brien and G.J. Chader. 1984. Synthesis of interphotoreceptor retinoid-binding protein (IRBP) by monkey retina in organ culture: Effect of monensin. Biochem. Biophys. Res. Commun. 118:789-796.

Yoshikami, S. and G.N. Noll. 1978. Isolated retinas synthesize visual pigments from retinol congeners delivered by liposomes. Science 200:1393-1395.

Yoshizawa, T. and Y. Shichida. 1982. Low-temperature spectrophotometry of intermediates of rhodopsin. Methods in Enzymology 81:333-354.

Zimmerman, W.F., R. Lion, F.J.M. Daemen and S.L. Bonting. 1975. Distribution of stereospecific retinol dehydrogenase activities in sub-cellular fractions of bovine retina and pigment epithelium. Exp. Eye Res. 21:325-332.

Vitamin A and Lipofuscin

W. Gerald Robison, Jr. and Martin L. Katz

Vitamin A, vitamin E, and phospholipids containing polyunsaturated fatty acids play important roles in both the structure and function of the retina, especially with regard to the photoreceptor cells (Robison et al., 1982). Deficiencies in dietary vitamin A, in forms which can be utilized as retinol, lead to night blindness, loss of photoreceptor outer segments, irreversible loss of photoreceptor nuclei, and ultimately to complete and permanent blindness (Dowling and Wald, 1958; 1960; Dowling and Gibbons, 1961; Wald, 1968; Carter-Dawson et al., 1979). The photosensitive portions (outer segments) of rod photoreceptor cells consist of densely arrayed sheets of membranes in the form of stacks of tightly packed discs which are specialized for visual transduction (Bonting et al., 1977; Rosenkranz, 1977). Approximately 90% of the protein in these membrane discs is opsin which, combined with a chromophoric group (11-cis retinaldehyde), constitutes the visual pigment, rhodopsin (Wald, 1968). Apparently, the rhodopsin is such an essential structural component of the membranes that in the absence of sufficient vitamin A to form this pigment, the very integrity of the membrane structure is compromised. Vitamin E deficiency also results in disruption of outer segment disc membranes and eventual loss of photoreceptor nuclei (Malatesta, 1951; Hayes et al., 1970; Hayes, 1974a; 1974b; Katz et al., 1978; 1982; 1984a; 1986a; Robison et al., 1979; 1980; 1982; Amemiya, 1981; Riis et al., 1981). In this instance, the loss of membrane integrity probably results from increased autoxidation of membrane lipids in the absence of the antioxidant activity of vitamin E. The outer segment membranes contain an unusually high proportion of highly unsaturated fatty acids in their phospholipids which make them extraordinarily susceptible to damage by autoxidation. Evidence regarding the importance of antioxidants, including vitamin E, in the maintenance of photoreceptor membranes, and studies of the contribution of autoxidized membrane products to the formation of lipofuscin (aging pigment) in the retinal pigment epithelium will be reviewed. Recent findings which implicated vitamin A as well as aging, vitamin E, and unsaturated fatty acids in determining the rates of formation and possible contents of lipofuscin granules in the retinal pigment epithelium will be considered. The implications of these findings to improved diagnosis and treatment of visual disorders will be discussed.

AGING AND RETINAL LIPOFUSCIN

A major difference between retinas from young and old individuals is the greater lipofuscin content in old retinas (Feeney et al., 1965; Feeney, 1978; Feeney-Burns et al., 1980; 1984; Wing et al., 1978; Katz and Robison, 1984). This lipofuscin is not located in the neuron cell bodies, or in any other part of the neural retina, as might be predicted from aging studies of the spinal cord and brain (Mann and Yates, 1974; Brizzee and Ordy, 1981). It is located, instead, in the retinal pigment epithelium (figures 1 and 2), which is a single layer of cells strategically located between the nutrient-laden vessels of the choriocapillaris and the photoreceptor cells (Zinn and Marmor, 1979). Developing an understanding of why the retinal pigment epithelium is essentially the only repository for lipofuscin in the entire retina would be of great interest. It is relevant to note that the retinal pigment epithelium is dynamically involved in fluid transport, and plays a major role in vitamin A transport and storage in the eye (Popper, 1944; Wald, 1968; Bok and Heller, 1976; Hirosawa and Yamada, 1976; Robison and Kuwabara, 1977; Young and Bok, 1979; Bok, 1985). At the same time, the retinal pigment epithelium apparently experiences relatively high oxygen fluxes (Sickel, 1972; Young and Bok, 1979), since most of the supplies required by photoreceptor cell activity pass through its cytoplasm (Dollery et al., 1969; Alm and Bill, 1972; Zuckerman and Weiter, 1980; Alder et al., 1983).

Quantitative measurements (figure 3) show a progressive age-related increase in the amount of lipofuscin in the retinal pigment epithelium (Wing et al., 1978; Katz and Robison, 1984). The

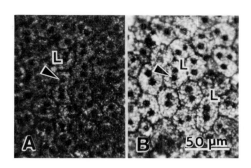

Figure 1. Lipofuscin autofluorescence in 4-month-old (A) and 30-month-old (B) ACI rats seen in unfixed flat preparations of retinal pigment epithelium. Regions containing lipofuscin granules (L) are so plentiful at 30 months that only the nuclear regions (arrowheads) and cell boundaries of these generally binucleate cells remain dark.

increase with time is similar to findings in other tissues (Porta and Hartroft, 1969; Miquel et al., 1977; Sohal, 1981), but the amount of lipofuscin which accumulates in the retinal pigment epithelium during aging, relative to tissue volume, is much greater than that found in most other tissues (Brizzee and Ordy, 1981; Katz et al., 1984b). Notable is the fact that the cilliary pigment epithilium, which has a similar developmental origin and is continuous with the retinal pigment epithelium at the ora serrata, shows no such lipofuscin accumulation either during aging or as a result of vitamin E deficiency (Robison et al., 1982). Although somewhat structurally similar to the retinal pigment epithelium, the cilliary pigment epithelium is not involved in maintenance of the photoreceptor cells, in significant levels of membrane ingestion by phagocytosis, in vitamin A transport and storage, nor does it contain unusual numbers of microperoxisomes as does the retinal pigment epithelium (Robison et al., 1982).

Abundant evidence suggests that lipofuscin is formed at least partially as a result of lipid autoxidation (Miquel et al., 1977; Katz et al., 1984a; Katz and Robison, 1986). Products of lipid autoxidation apparently accumulate intracellularly with time, and particularly rapidly in cells where there is substantial turnover of lipids which can undergo autoxidation reactions. A variety of evidence suggests that malonaldehyde, formed by autoxidation of polyunsaturated fatty acids, can combine with free amino groups of various compounds, and form autofluorescent lipopigment complexes (Tappel et al., 1973; Tappel, 1975; Elleder, 1981; Siakotos and Munkres, 1981). Recent evidence, however, suggests that other mechanisms may be involved in lipofuscin fluorophore formation in the retinal pigment epithelium (Eldred et al., 1982; Eldred, 1986).

Certain metabolic characteristics of the retinal pigment epithelium suggest that the autoxidation of polyunsaturated fatty acids may contribute to lipofuscin formation in this tissue. The retinal pigment epithelium regularly phagocytizes membrane discs discarded from the rod outer segments (LaVail, 1976). The membranes are shed into the subretinal space in packets of 8 to 30 discs (Young and Bok, 1969; Bok and Young, 1979). The packets are ingested in phagosomes which fuse with primary lysosomes and undergo further fusions and changes as their contents are broken down, processed and/or recycled by the phagolysosomal system of the retinal pigment epithelium (Feeney-Burns and Eldred, 1983). The disc membranes have an unusually high proportion of long-chain, polyunsaturated fatty acids and a low cholesterol content (Daemen, 1973; Anderson et al., 1977; 1978; Fliesler and Anderson, 1983). Usually about 50% and sometimes as much as 90% of the fatty acids in these photoreceptor membranes are polyunsaturates. Apparently, the high degree of unsaturation is critical for providing high fluidity and other characteristics of the microenvironment required for the normal function of rhodopsin and stability of membrane structure (Bonting et al., 1977; Robison et al., 1982). The high level of unsaturation of the photoreceptor membrane fatty acids occurs in many species and is maintained under various environmental and dietary conditions (Anderson, 1970). Specific long-chain unsaturated fatty acids are retained in the outer segments even with dietary deficiency in the essential fatty acid precursors and depletion of polyunsaturates from most body tissues (Dudley et al., 1975; Anderson et al., 1976a). Apparently, the products generated by autoxidation of polyunsaturated fatty acids are highly resistant

Figure 2. Transections of retinal pigment epithelium from 4-month-old (A), 11-month-old (B), and 32-month-old (C) ACI rats showing the relative numbers of lipofuscin granules (L), phagosomes (P), melanin granules or premelanosomes (M), and primary or secondary lysosomes (arrowheads) within this single layer of cells which is located between the rod outer segments (ROS) of the neural retina and the capillaries (C) of the choriocapillaris.

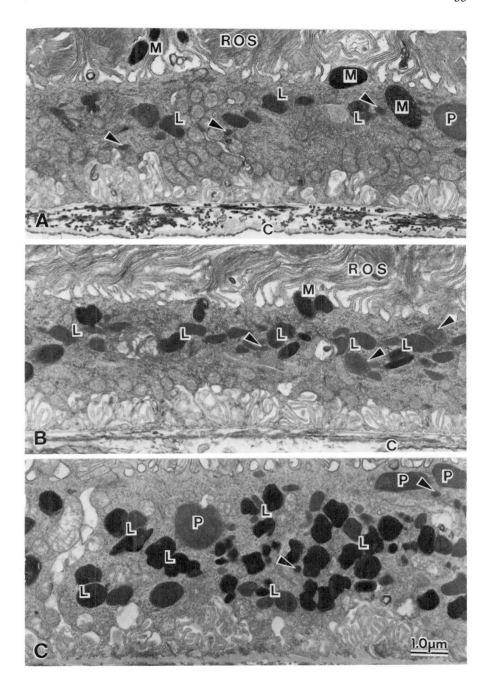

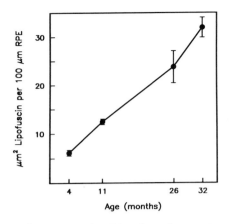

Figure 3. Effect of age on the lipofuscin content of the retinal pigment epithelium in pigmented rats. The amount of lipofuscin increased almost linearly with age. At 32 months, the lipofuscin content of the retinal pigment epithelium was over five times that at 4 months ($\underline{P} < 0.001$). Adapted from Katz and Robison (1984) with permission.

to degradation by the enzymes of intracellular phagolysosomal systems, and to extraction by organic solvents, thus satisfying the classical definition of lipofuscin (Porta and Hartroft, 1969; Pearse, 1972; Elleder, 1981).

Since a single retinal pigment epithelial cell may ingest and process as many as 30,000 disc membranes every day (Bok and Young, 1979), it seems reasonable to propose that the massive accumulations of lipofuscin in the retinal pigment epithelium may represent, at least in part, autoxidation products derived from photoreceptor membranes. Besides unsaturated lipids, these membranes also contain vitamin A which likewise is very susceptible to autoxidation and might contribute to lipofuscin formation.

If lipofuscin does indeed represent products of autoxidation, then manipulating the antioxidant content of tissues would be expected to alter the rates of lipofuscin formation within them. The retinal pigment epithelium should be an ideal tissue in which to examine the effects of autoxidation on lipofuscin formation. This tissue normally accumulates great amounts of lipofuscin, and is directly involved in a dynamic turnover of many polyunsaturated fatty acids and vitamin A in an environment where there is a high oxygen flux.

VITAMIN E DEFICIENCY, LIPIDS, AND RETINAL LIPOFUSCIN

Dietary deficiency in vitamin E, with or without deficiencies in various other antioxidants, does in fact significantly accelerate lipofuscin accumulation in the retinal pigment epithelium (Katz et al., 1978; 1984a; 1986a; Robison et al., 1979; 1980; 1982). This accelerated deposition of lipofuscin appears to be a good model for retinal aging studies, since the vitamin E-deficiency-related lipofuscin accumulates in the same place (in the retinal pigment epithelium) as does age-related lipofuscin, and it also exhibits the same fluorescence emission spectrum (Katz et al., 1984b).

Vitamin E is an antioxidant known to protect biological membrane lipids from autoxidation, and it is a consistent component of rod outer segments of animals on normal diets (Dilley and McConnell, 1970; Farnsworth and Dratz, 1976; Hunt et al., 1984). Deficient diets result in tissue deficiencies, disruption of the oxygen-labile outer segment membranes, and marked acceleration in the accumulations of lipofuscin in the retinal pigment epithelium. Monkeys deficient in vitamin E for 2 years showed degeneration of photoreceptor outer segments and unusually high lipofuscin accumulation in the retinal pigment epithelium of the macular region (Hayes, 1974a). Katz et al. (1978) found massive accumulations of lipofuscin in the pigment epithelium of pigmented rats fed diets lacking various antioxidants, including vitamin E. Robison et al. (1979; 1980; 1982) demonstrated ultrastructural disruption of rod outer segment membrane discs and progressive loss of photoreceptor nuclei, as well as marked increases in lipofuscin deposits in the retinal pigment epithelium of albino rats fed vitamin E-deficient diets for 20 to 34 weeks (figure 4). A fivefold acceleration of lipofuscin accumulation was accompanied by the loss of structural integrity of the outer segments. The retinal damage resulting from vitamin E deficiency commenced in the oxidation-labile outer segment membranes and led to photoreceptor death, but left the remainder of the retina structurally normal. This evidence suggests that the lipofuscin formed represents the products of autoxidation of photoreceptor membrane components which are ingested by the retinal pigment epithelium.

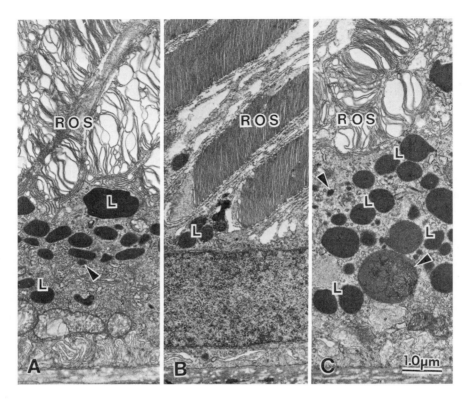

Figure 4. The effect of vitamin E deficiency on the retina. Transections of the retinal pigment epithelium and portions of the rod outer segments (ROS) from rats that were fed a normal diet for 8 months (B), and from rats that were fed a vitamin E-free diet for 5 months (A) and 8 months (C). The relative numbers of lipofuscin granules (L) and lysosomes (arrowheads) are seen. Reproduced from Robison et al., 1979 with permission of the Association for Research in Vision and Ophthalmology.

PHOTORECEPTOR CELLS AND RETINAL LIPOFUSCIN

If daily ingestion and processing of photoreceptor outer segment membranes does, in fact, contribute significantly to the deposition of lipofuscin in the retinal pigment epithelium, then decreasing or eliminating phagocytosis of discarded portions of outer segments should decrease the accumulation of lipofuscin. Therefore, studies of lipofuscin formation were carried out in animals which had no photoreceptor outer segment material for most of their lives, and thus did not discard membrane discs for daily ingestion by the retinal pigment epithelium. Rats with the gene for retinal dystrophy (rdy) begin to show degeneration of photoreceptor outer segments by 12 days after birth and they lose most of the photoreceptor nuclei by 60 to 70 days of age, leaving only cellular debris in the subretinal space (LaVail, 1979). There is little, if any, normal recycling of rod outer segment membranes with the involvement of the retinal pigment epithelium in lipid processing. In fact, phagocytosis by the retinal pigment epithelium is minimal (LaVail, 1979).

Rats homozygous for the rdy gene (RCS-p/p,rdy/rdy) and their congenic controls (RCS-p/p,+/+) were used to evaluate the possible contribution of photoreceptor cells to lipofuscin deposition in the retinal pigment epithelium (Robison et al., 1982; Katz et al., 1986b). At 66 days of age, both groups of rats had similar amounts of lipofuscin in the retinal pigment epithelium, but by 16 months the animals with normal retinas had accumulated substantially more lipofuscin than did those with retinas lacking photoreceptors (figure 5). This finding suggests that photoreceptor cells play a significant role in lipofuscin deposition in the retinal pigment epithelium. Evidence that components of the photoreceptor outer segments can be converted directly into lipofuscin fluorophores is provided by the recent discovery that, in young rats with retinal dystrophy, the degenerating remnants of photoreceptor cell membranes have fluorescence properties similar to those of lipofuscin in the retinal pigment epithelium (Katz et al., 1986b).

Another model used to evaluate the possible involvement of photoreceptor cells in lipofuscin formation was the albino rat in which most of the photoreceptor cells had been lost as a result of retinal light damage. Albino rats were exposed to continuous illumination of 400 foot-candles for 6 weeks in order to eliminate almost all of their photoreceptor cells. The age-related accumulation of lipofuscin in the retinal pigment epithelium of these light-damaged rats and of normal controls were compared (Robison et al., 1982). The retinas of the light-exposed rats had essentially no complete photoreceptor cells and only zero to one row of photoreceptor nuclei, whereas the control animals had complete photoreceptor cells and 9 to 11 rows of photoreceptor nuclei. The rats lacking rod outer segments due to light damage provided a good experimental model for defining the role of outer segment phagocytosis in lipofuscin deposition because they had an apparently

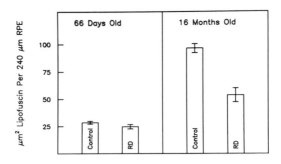

Figure 5. Effect of hereditary photoreceptor degeneration (<u>rdy</u> gene) on the lipofuscin content of the retinal pigment epithelium. The age-related accumulation of lipofuscin in the retinal pigment epithelium of the control group was significantly greater than that in the dystrophic group ($P < 0.01$). Reproduced from Katz et al., 1986b with permission from Exp. Eye Res.

normal pigment epithelium yet still lacked a daily source of lipid-rich disc membranes. As with the genetically blind rats, the light-blinded rats accumulated substantially less lipofuscin during aging than did the normal rats used as controls (figure 6A-6C). Apparently, in the normal retina, the regular phagocytosis and processing of photoreceptor outer segment membranes by the retinal pigment epithelium is related to lipofuscin deposition in a way that suggests that perhaps the membranes themselves contribute a structural component to the lipofuscin formed.

VITAMIN A AND RETINAL LIPOFUSCIN

Vitamin A, like the lipids discussed above, is at times bound in the membranes of the outer segments, and at other times is being processed or stored in the retinal pigment epithelium. Apparently,

Figure 6. Lipofuscin (L) in the retinal pigment epithelium of dystrophic (A & D), normal (B & E), and light-damaged (C & F) retinas of albino rats fed a vitamin A-supplemented diet (D+A, N+A, & LD+A, respectively) or a vitamin A-free diet (D-A, N-A, & LD-A, respectively) for 9 months from weaning. The lipofuscin granules (L) are seen in methacrylate (1 μm) sections (above) and the lipofuscin autofluorescence is shown in frozen (12 μm) sections (below) for each group. The retina with normal photoreceptor cells and intact rod outer segments (ROS) exhibited the most lipofuscin, but all the retinas from rats receiving vitamin A had markedly more lipofuscin than did the retinas from vitamin A-deprived rats.

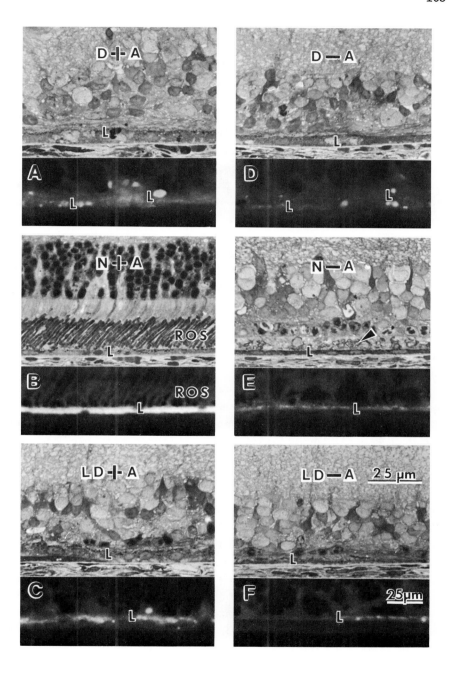

exchange of vitamin A between the photoreceptor cells and the retinal pigment epithelium occurs continually (Defoe and Bok, 1983). However, in the rat eye, during the dark phase of a dark/light cycle, probably most of the vitamin A is localized in the visual pigment of the outer segments in the form of 11-cis retinaldehyde bound to opsin (Wald, 1968; Young, 1971). Upon light onset, the 11-cis retinaldehyde is isomerized, released from opsin, and reduced, so that all-trans retinol is freed in large quantities and is transfered very rapidly to the retinal pigment epithelium where it is esterified (mainly as retinyl palmitate and retinyl stearate) and stored (Schairer and Patzelt, 1940; Dowling and Wald, 1958; 1960; Dowling and Gibbons, 1961; Wald, 1968; Bridges, 1976; Bridges et al., 1982). Upon dark onset the vitamin A esters are hydrolyzed and vitamin A is transported to the photoreceptor cell membranes and is converted to 11-cis retinaldehyde, which is incorporated into visual pigment, thus completing the cycle. During a single dark/light cycle of this dynamic exchange, much of the vitamin A of the eye passes through the retinal pigment epithelium. It is clear that a major function of the retinal pigment epithelium which distinguishes it from other eye tissues is a rapid handling and storage of large quantities of vitamin A.

Our studies on vitamin E-deficient rats were the first to establish a clear effect of vitamin A (retinol) on the lipofuscin content of the retinal pigment epithelium (Robison et al., 1980), although several investigators had previously suggested the idea that retinol or retinaldehyde should be susceptible to oxidation in the retina (Daemen, 1973; Hayes, 1974a). As described above, rats fed vitamin E-deficient diets exhibited a marked increase in the amount of lipofuscin which accumulated in the retinal pigment epithelium (Robison et al., 1979), but the effect of vitamin E deficiency was greatly influenced by varying the amounts of vitamin A (retinol) available as dietary retinyl acetate or retinyl palmitate. Rats fed vitamin E-deficient diets containing either retinol (-E,+A) or retinoic acid (-E,-A) were compared to controls fed diets supplemented with vitamin E and containing either retinol (+E,+A), or retinoic acid (+E,-A). Although most body tissues are able to utilize retinoic acid when provided in the diet, the retina requires retinol for rhodopsin synthesis (Dowling and Wald, 1958; 1960; Dowling and Gibbons, 1961; Wald, 1968). In rats restricted to retinoic acid as their source of vitamin A, not only was there a predictable loss of outer segment integrity and dropout of photoreceptor nuclei, but there was a striking decrease in the amount of lipofuscin which would be expected to accumulate in the retinal pigment epithelium in vitamin E-deficient rats, as measured by either autofluorescence or by the number of lipofuscin granules. Rats of the −E, −A group (retinol-deprived) showed only a twofold increase in numbers of lipofuscin granules, whereas the −E, +A rats displayed the expected fivefold increase over that found in the retinas of control rats (+E, +A).

In contrast to retinal lipofuscin, the lipofuscin of the uterus was not affected by changes in dietary vitamin A (Bieri et al., 1980). A cursory interpretation of the evidence might suggest that the effect of vitamin A on lipofuscin deposition is retina-specific, but this is not necessarily true. It is probably only because of the retina's dependence on the retinol form of vitamin A (Dowling and Gibbons, 1961) that the experiments cited were able to demonstrate a retinol-effect on lipofuscin deposition. Apparently, the uterus contains binding proteins for both retinoic acid and retinol (Chytil et al., 1975), whereas the retinal pigment epithelium may have only a retinol binding protein (Wiggert et al., 1979). Since the uterus can utilize retinoic acid, as can most body tissues outside the eye, the animals in all the experimental groups listed (even the +E,−A and −E,−A groups) had adequate vitamin A in the form of either retinoic acid or retinol, so no group had a truly vitamin A-deprived uterus. Additional experiments need to be designed utilizing low levels of retinoic acid or some other method to make all forms of vitamin A unavailable or less available for possible influences on lipofuscin, before it can be concluded that the effects of vitamin A on lipofuscin are limited to the retina. In fact, there is some evidence for an effect of vitamin A on the lipofuscin content of extraocular muscle (Herman et al., 1985) and of the choroid (Herrmann et al., 1984).

An experiment was designed to determine if the vitamin A effect on the lipofuscin of the retinal pigment epithelium was dependent on the presence of normal photoreceptor cells (Robison et al., 1982). Rats missing all photoreceptor outer segments due to heredity (rdy gene) and non-dystrophic rats of a congenic stain, missing most of their rods due to light damage, were placed on diets either deficient or adequate in retinol and compared to similarly fed congenic rats with the normal gene and no retinal light damage.

Thus, rats with completely normal retinas, rats with retinas missing rod outer segments for ingestion but having a normal pigment epithelium, and rats having retinas with an essentially non-ingesting retinal pigment epithelium and no outer segments to ingest were available. This permitted comparisons of the relative contributions to lipofuscin formation by: 1) ingested outer segment membranes which contained polyunsaturated lipids plus vitamin A; and 2) the diet-related stores of vitamin A in the retinal pigment epithelium. Within each diet group, the retinal pigment epithelium of both rat groups which had no outer segments for ingestion exhibited somewhat less lipofuscin than did the retinal pigment epithelium of rats with normal outer segments and normal daily ingestion of their membranes (figure 6). As shown in experiments described above, the presence of photoreceptors and the regular phagocytosis of their outer segment membranes by the retinal pigment epithelium contribute to lipofuscin accumulation. The lipofuscin fluorescence could include contributions from the autoxidation products of either or both the polyunsaturated lipids and the vitamin A of the photoreceptor membranes.

The presence or absence of dietary vitamin A (retinol) had a much more striking effect on lipofuscin formation in the retinal pigment epithelium than did the presence or absence of photoreceptor cells (compare figures 6A-C with 6D-F). Between diet groups, the lipofuscin content was always markedly lower in vitamin A-deficient retinas. Even in retinas lacking photoreceptor cells, vitamin A deprivation resulted in greatly reduced amounts of lipofuscin. Therefore, the presence or absence of retinol has a greater influence than does a source of outer segment membranes. Apparently, vitamin A derived from sources other than the photoreceptor cells could be a major substrate for lipofuscin formation in the retinal pigment epithelium. Although phagocytosis by the retinal pigment epithelium is known to decrease in vitamin A-deficient animals (Dowling and Gibbons, 1961; Herron and Riegel, 1974), the vitamin effect could not be explained by a decrease in phagocytosis alone, since it occurred even when no photoreceptor components remained to be ingested. Where no photoreceptor cells were present, the vitamin A effect might be explained in part by a reduction in autophagy by the retinal pigment epithelium in the vitamin A-deficient rats (Dowling and Gibbons, 1961; Reme, 1977; Young and Bok, 1979) and other changes resulting in less turnover of lipids. However, the influence of these factors is probably minor, since the lipids endogenous to the retinal pigment epithelium are mainly saturated (Anderson et al., 1976b; Stone et al., 1979; Braunagel, et al., 1985) and would not be expected to contribute significantly to lipofuscin deposition. It is more likely that the retinol effect is mainly the result of direct incorporation of vitamin A autoxidation products into the lipofuscin of the retinal pigment epithelium.

Recent evidence supports a direct involvement of vitamin A (retinol or retinaldehyde) in the formation and/or accumulation of lipofuscin. Retinas of rats with the retinal dystropy gene (rdy) fail to recycle the disc membranes of photoreceptor outer segments. The outer segments continue to elongate and the discarded portions of their distal ends accumulate as whorls of membrane debris in the space between the outer segments and the retinal pigment epithelium (LaVail et al., 1972; LaVail, 1979). It was found that this membrane debris had an autofluorescence with a yellow emission like that of the lipofuscin in the adjacent retinal pigment epithelium (Katz et al., 1986b). Since it is clear that the debris came from discarded outer segment membranes, the resulting autofluorescence strongly suggests that disc membrane components such as polyunsaturated lipids and/or vitamin A became autoxidized (Zigler and Hess, 1985) and provided the fluorescent character of the debris. This suggests that the lipofuscin fluorophores of the retina probably derive from outer segment components in normal retinas.

In retinal dystrophic rats fed a retinol-free diet for 7 weeks from weaning, the autofluorescence of the debris was markedly decreased from that observed in control rats fed a retinol-supplemented diet (figure 7). This would appear to implicate vitamin A directly in the process of lipofuscin formation. It may be that the vitamin A itself, while located in the photoreceptor membranes, undergoes autoxidation and becomes an autofluorescent component of the debris.

Since vitamin A appears to be associated directly with lipofuscin formation, different dietary levels of retinol would be expected to result in different degrees of lipofuscin accumulation in the retinal pigment epithelium, providing the different levels chosen resulted in different amounts of vitamin A being available in this tissue. A series of rats were fed diets either adequate (+E)

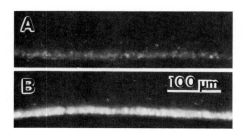

Figure. 7 Lipofuscin-like autofluorescence of the debris that accumulated from degenerating photoreceptor cells in the retinas of rats with the retinal dystrophic gene (rdy) which were fed a vitamin A-free diet (A) or a vitamin A-supplemented diet (B) for 46 days from weaning. From Katz et al., in preparation.

or deficient (-E) in vitamin E, and containing very different level of vitamin A, some at truly marginal levels (Robison et al., 1982). Both +E and -E groups of rats received 23.0 mg; 0.23 mg; 0.058 mg or 0.00 mg; retinol/kg diet from weaning. Retinoic acid (4 mg/kg diet) was fed to the .058 mg and retinol-deprived groups in order to maintain the nonretinal tissues of the body in a healthy state.

After 18 weeks and 30 weeks on the diets, the retinal pigment epithelial cells of all the vitamin E-deficient rats exhibited much more lipofuscin autofluorescence in frozen sections and many more lipofuscin granules in plastic sections than did those of any vitamin E-adequate rats (figures 8 and 9). Within the vitamin E diet groups, the rats receiving increasing amounts of dietary vitamin A had increasingly more lipofuscin in their retinal pigment epithelium. This dose-related response suggests a close relationship between dietary vitamin A and lipofuscin formation.

Further studies were undertaken to determine if there would be a different retinol effect between rats fed a fully adequate amount of vitamin A (4 mg retinol/kg diet) and those fed an excessive level (23 mg retinol/kg diet) compared to rats deprived of vitamin A (Katz et al., 1986a; 1986c). Updated methods of ultrastructural morphometry utilizing computer planimetry were employed in this

Figure 8. Transections of retinal pigment epithelium showing the relative numbers of lipofuscin granules (L), phagosomes (P), and primary or secondary lysosomes (arrowheads) from Sprague-Dawley rats which were fed diets containing adequate vitamin E and different amounts of vitamin A: (A) 23 mg/Kg diet; (B) 0.23 mg/Kg diet; (C) 0.058 mg/Kg diet; and (D) 0.00 mg/Kg diet for 8 months from weaning.

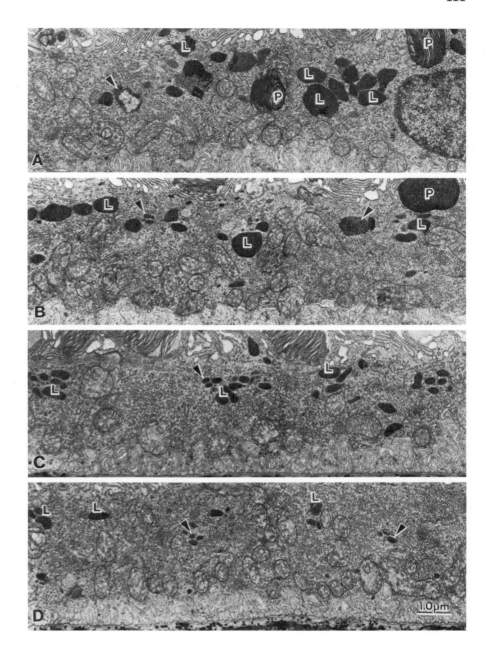

study.   There appeared to be no significant differences in amounts
of lipofuscin between rats fed adequate and excessive levels of
vitamin A.   Measurements of the vitamin A storage levels in the
retinal pigment epithelium were consistent with these findings.
They showed the retinyl esters to be at essentially normal levels in
rats receiving either the adequate or excessive levels of vitamin A
(figure 10).   Thus, the lipofuscin content of the retinal pigment
epithelium appeared to be directly related to the levels of retinyl
esters in this tissue. Apparently, only dietary levels at or less
than 0.23 mg retinol/kg diet are effective in lowering the tissue
levels of vitamin A sufficiently to cause a decrease in vitamin A-
related lipofuscin deposition, while excessive dietary vitamin A
levels do not result in either elevated vitamin A content or
increased lipofuscin accumulation in the retinal pigment epithelium.

LIPOFUSCIN AND VISUAL DISORDERS

The retinal pigment epithelium plays a dynamic metabolic role
in the maintenance of photoreceptor cells, and is the almost
exclusive site of lipofuscin deposition in the retina.   Since many
visual disorders appear to result from primary changes in the
retinal pigment epithelium (Hogan, 1972; Mullen and LaVail, 1976;
Marmor, 1979), a better understanding of the mechanisms and effects
of lipofuscin deposition in this tissue may aid in the diagnosis and
treatment of retinal diseases.   Retinal lipofuscin may be an
indicator of tissue autoxidative damage, failure of antioxidant
systems, levels of tissue retinol, and/or various disease states.
Retinitis pigmentosa (Kolb and Gouras, 1974; Szamier et al.,
1979), ceroid lipofuscinosis (Zeman, 1971; Neville et al., 1980;
Armstrong and Koppang, 1981), fundus flavimaculatus (Eagle et al.,
1980), and abetalipoproteinemia (Cogan et al., 1984), exhibit
ocular changes which involve degeneration of photoreceptor cells and
accumulation of lipofuscin in the retinal pigment epithelium.
Batten's disease, one of the ceroid lipofuscinoses, reportedly

Figure 9.   Transection of retinal pigment epithelium showing the
relative numbers of lipofuscin granules (L), phagosomes (P), and
primary or secondary lysosomes (arrowheads) from Sprague-Dawley rats
which were fed diets lacking vitamin E but containing different
amounts of vitamin A: (A) 23 mg/Kg diet; (B) 0.23 mg/Kg diet; (C)
0.058 mg/Kg diet; and (D) 0.00 mg/Kg diet for 8 months from weaning.

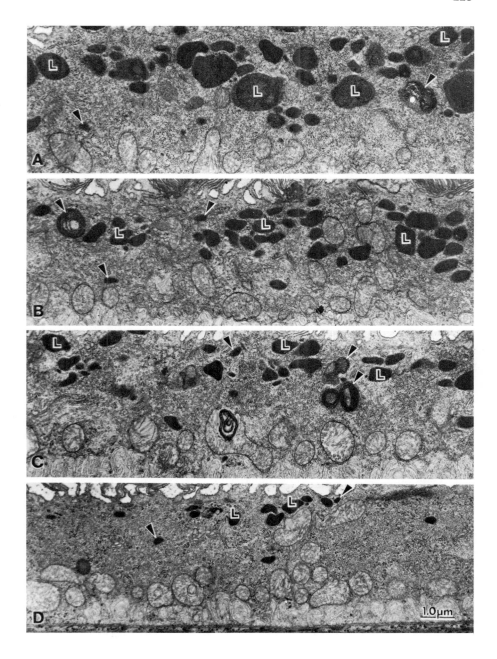

114

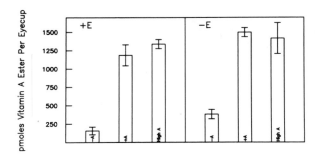

Figure 10. Relationship between dietary vitamins A and E and the total retinyl ester content of the retinal pigment epithelium in light-adapted Fisher 344 albino rat eyes. The +A group received 8.0 mg/Kg vitamin A in the diet and the high A group received 23.0 mg/Kg diet. From Katz et al. (1986c).

involves changes in peroxidative mechanisms (Armstrong et al.,1974). Extracts from the massive lipofuscin deposits in the brain of a patient with Batten's disease have been reported to contain derivatives of vitamin A (retinoyl compounds: Wolfe et al., 1977). However, just how lipofuscin and the factors which influence its formation are related to ocular disease is not known.

The types of lipofuscin which accumulate as a result of various diseases may be different from the age-related, vitamin E-deficiency-related, or vitamin A-related lipofuscins. If so, any consistent differences identified could be used as criteria for distinguishing origins of lipofuscin types. One method which may be useful for determining differences is spectral analysis of the fluorescence emission using equipment corrected for wavelength sensitivity. Also, the development of new methods of extraction and chemical analysis could be pursued. If reliable methods were established for differentiating lipofuscins according to the origin of their contents, these could serve as powerful diagnostic tools for various physiological and disease states.

SUMMARY AND CONCLUSIONS

Retinal lipofuscin is deposited almost exclusively in the retinal pigment epithelium, where it appears to represent end products of metabolic activity. It accumulates progressively as the result of normal aging, or at greatly altered rates as the result of dietary changes and certain disease states. The retinal pigment epithelium is unusual in that it is a post-mitotic tissue which ingests, processes, and transports extraordinary amounts of unsaturated lipid and vitamin A, both of which are very susceptible to oxidation, and does this in an environment where there is a relatively high oxygen flux. At the same time, the retinal pigment epithelium exhibits striking lipofuscin accumulations which are clearly influenced by increased age, antioxidant deficiencies, and/or vitamin A levels. A variety of experimental evidence is consistent with the possibility that the lipofuscin of the retinal pigment epithelium represents autoxidation products of both phospholipids and vitamin A.

Vitamin A, in the form of retinaldehyde, and unsaturated phospholipids are major components of the photoreceptor disc membranes which are ingested daily by the retinal pigment epithelium. Remnants of disc membranes which are discarded in the subretinal space of rats with retinal dystrophy exhibit autofluorescence similar to that of lipofuscin, suggesting that components of the photoreceptor outer segments are probably precursors of lipofuscin fluorophores. Further evidence that derivatives of phospholipids and/or vitamin A are involved in liposfuscin formation comes from studies of retinas which have no photoreceptor outer segments to be ingested by the retinal pigment epithelium. These exhibit decreased accumulations of lipofuscin, as would be expected if products of outer segment membranes were important for lipofuscin deposition.

Experimental evidence suggests that the lipofuscin of the retinal pigment epithelium is formed at least in part from vitamin A, possibly retinaldehyde of the outer segments and/or retinyl esters which are stored in the retinal pigment epithelium during the light cycle. Lack of dietary vitamin A results in a striking decrease in lipofuscin accumulation both in rats having normal retinas and in rats having retinas without photoreceptor cells, compared to the respective controls fed vitamin A.

One possible mechanism by which vitamin A may promote lipofuscin formation in the retinal pigment epithelium is by acting as a photosensitizer for autoxidation of molecular components of the photoreceptor cells or of the retinal pigment epithelium. Apparently, illumination of retinaldehyde can result in the production of singlet oxygen (Delmelle, 1978), a species highly reactive with biological membranes due to its strong electrophilic properties that direct it towards electron-rich chemical bonds such as those found in unsaturated lipids. Upon reaction of singlet oxygen with unsaturated fatty acids, lipid peroxides are formed which are broken down to form free radical products capable of initiating free radical chain reactions with surrounding unsaturated fatty acid molecules and possibly with vitamin A species. A decrease in the levels of docosahexaenoic acid, the major polyunsaturated fatty acid in rod outer segments, and an apparent increase in lipid hydroperoxides was observed following 24 to 72 hours of constant light exposure (Wiegand et al., 1983). Apparently, the autoxidation of long-chain polyunsaturated fatty acids results in the formation of autofluorescent products which are deposited in the retinal pigment epithelium as residual, cellular inclusions defined as lipofuscin. The facts that vitamin A deficiency protects against retinal light damage (Noell et al., 1971; Carter-Dawson et al., 1981) and also results in decreased lipofuscin accumulation in the retinal pigment epithelium (Robison et al., 1982) are consistent with an involvement of vitamin A in the peroxidation of retinal membrane components and their inclusion in lipofuscin deposits.

If products of vitamin A autoxidation are themselves included in lipofuscin deposits, one should be able to detect some differences in the properties of lipofuscin formed in animals fed vitamin A-adequate versus vitamin A-deficient diets. Experiments are underway currently to determine the effects of dietary vitamin A on the spectral properties and chemical compositon of lipofuscin of the retinal pigment epithelium. As better methods are developed for analysis of possible differences in types of lipofuscin according to its origin, these could be useful in the diagnosis of lipofuscin-related ocular diseases.

ACKNOWLEDGEMENTS

We wish to thank Dr. John G. Bieri, Mrs. Anne B. Groome, and Miss Christine M. Drea for their expert assistance.

REFERENCES

Alder, V.A., S.J. Cringle and I.J. Constable. 1983. The retinal oxygen profile in cats. Invest. Ophthalmol. Vis. Sci. 24:30-36.

Alm, A. and A. Bill. 1972. The oxygen supply to the retina. II. Effects of high intraocular pressure and of increased arterial carbon dioxide tension on uveal and retinal blood flow in cats. Acta Physiolog. Scand. 84:306-319.

Amemiya, T. 1981. Photoreceptor outer segment and retinal pigment epithelium in vitamin E deficient rats. Graefs Archiv Ophthalmologie 216:103-109.

Anderson, R.E. 1970. Lipids of ocular tissues IV. A comparison of the phospholipids from the retina of six mammalian species. Exp. Eye Res. 10:339-344.

Anderson, R.E., R.M. Benolken, M.B. Jackson and M.B. Maude. 1977. The relationship between membrane fatty acids and the development of the rat retina. Advan. Exp. Med. Biol. 83:547-559.

Anderson, R.E., R.M. Benolken, P.A. Kelleher, M.B. Maude and R.D. Wiegand. 1978. Chemistry of photoreceptor membrane preparations from squid retinas. Biochemica Biophysica Acta 510:316-326.

Anderson, R.E., D.J. Landis and P.A. Dudley. 1976a. Essential fatty acid deficiency and renewal of rod outer segments in the albino rat. Invest. Ophthalmol. Vis. Sci. 15:232-236.

Anderson, R.E., P.M. Lissandrello, M.B. Maude and M.T. Matthes. 1976b. Lipids of bovine retinal pigment epithelium. Exp. Eye Res. 23:149-157.

Armstrong, D., S. Dimmitt, D.E. Van Wormer and T. Okla. 1974. Studies in Batten Disease I. peroxidase deficiency in granulocytes. Arch. Neuro. 30:144-152.

Armstrong, D. and N. Koppang. 1981. Ceroid-lipofuscinosis, a model for aging. In: Age Pigments, R.S. Sohal, ed., Elsevier/North-Holland Biomedical Press, Amsterdam, pp.355-382.

Bieri, J.G., T.J. Tolliver, W.G. Robison, Jr and T. Kuwabara. 1980. Lipofuscin in vitamin E deficiency and the possible role of retinol. Lipids 15:10-13.

Bok, D. 1985. Retinal photoreceptor-pigment epithelium interactions. Invest. Ophthalmol. Vis. Sci. 26:1659-1694.

Bok, D. and J. Heller. 1976. Transport of retinol from the blood to the retina: an autoradiographic study of the pigment epithelial cell surface receptor for plasma retinol-binding protein. Exp. Eye Res. 22:395-402.

Bok, D. and R.W. Young. 1979. Phagocytic properties of the retinal pigment epithelium. In: The Retinal Pigment Epithelium, K.M. Zinn and M.F. Marmor, eds., Harvard University Press, Cambridge, Mass., pp. 148-174.

Bonting, S.L., P.J.G.M. van Breugel and F.J.M. Daemen. 1977. Influence of the lipid environment on the properties of rhodopsin in the photoreceptor membrane. Advan. Exp. Med. Biol. 83:175-189.

Braunagel, S.C., D.T. Organisciak and H. Wang. 1985. Isolation of plasma membranes from the bovine retinal pigment epithelium. Biochimica Biophysica Acta 813:183-194.

Bridges, C.D.B. 1976. vitamin A and the role of the pigment epithelium during bleaching and regeneration of rhodopsin in the frog eye. Exp. Eye Res. 22:435-455.

Bridges, C.D.B., R.A. Alvarez and S-L. Fong. 1982. Vitamin A in human eyes: amount, distribution, and composition. Invest. Ophthalmol. Vis. Sci. 22:706-714.

Brizzee, K.R. and J.M. Ordy. 1981. Cellular features, regional accumulation, and prospects of modification of age pigments in mammals. In: Age Pigments, R.S. Sohal, ed., Elsevier/North-Holland Biomedical Press, Amsterdam, pp. 102-154.

Carter-Dawson, L., T. Kuwabara and J.G. Bieri. 1981. Effects of moderate-intensity light on vitamin A-deficient rat retinas. Invest. Ophthalmol. Vis. Sci. 20:569-574.

Carter-Dawson, L., T. Kuwabara, P.J. O'Brien and J.G. Bieri. 1979. Structural and biochemical changes in vitamin A deficient rat retinas. Invest. Ophthalmol. Vis. Sci. 18:437-446.

Chytil, F., D.L. Page and D.E. Ong. 1975. Presence of cellular retinol and retinoic acid binding proteins in human uterus. Int. J. Vitam. Nutr. Res. 45:293-298.

Cogan, D.G., M. Rodrigues, F.C. Chu and E.J. Schaefer. 1984. Ocular abnormalities in abetalipoproteinemia, a clinicopathologic correlation. Ophthalmology 91:991-998.

Daemen, F.J.M. 1973. Vertebrate rod outer segment membranes. Biochimica Biophysica Acta 300:255-288.

Defoe, D.M. and D. Bok. 1983. Rhodopsin chromophore exchanges among opsin molecules in the dark. Invest. Ophthalmol. Vis. Sci. 24:1211-1226.

Delmelle, M. 1978. An investigation of retinal as a source of singlet oxygen. Photochem. Photobiol. 27:731-734.

Dilley, R.A. and D.G. McConnell. 1970. Alpha-tocopherol in the retinal outer segment of bovine eyes. J. Membrane Biol. 2:317-323.

Dollery, C.T., C.J. Bulpitt and E.M. Kohner. 1969. Oxygen supply to the retina from the retinal and choroidal circulation at normal and increased arterial oxygen tensions. Invest. Ophthalmol. Vis. Sci. 8:588-594.

Dowling, J.E. and I.R. Gibbons. 1961. The effect of vitamin A deficiency on the fine structure of the retina. In: The Structure of the Eye, G.K. Smelser, ed., Academic Press, New York, pp. 85-99.

Dowling, J.E. and G. Wald. 1958. Vitamin A deficiency and night blindness, Proc. Natl. Acad. Sci. (USA) 44:648-661.

Dowling, J.E. and G. Wald. 1960. The biological function of vitamin A acid. Proc. Natl. Acad. Sci. (USA) 46:587-608.

Dudley, P.A., D.J. Landis and R.E. Anderson. 1975. Further studies on the chemistry of photoreceptor membranes of rats fed an essential fatty acid deficient diet. Exp. Eye Res. 21:523-530.

Eagle, R.C., Jr., A.C. Lucier, V.B. Bernardino and M. Yanoff. 1980. Retinal pigment epithelial abnormalities in fundus flavimaculatus, a light and electron microscopic study. Ophthalmology 87: 1189-1200.

Eldred, G.E. 1986. Questioning the nature of the fluorophores in age pigments. In: Age Pigments: Biological Markers in Aging and Environmental Stress, E.A. Totaro, D. Armstrong, P. Glees and F.A. Pisanti, eds., NVU Science Press, Amsterdam (in press).

Eldred, G.E., G.V. Miller, W.S. Stark and L. Feeney-Burns. 1982. Lipofuscin: resolution of discrepant fluorescence data. Science 216:757-759.

Elleder, M. 1981. Chemical characterization of age pigments. In: Age Pigments, R.S. Sohal, ed., Elsevier/North-Holland Biomedical Press, Amsterdam, pp. 203-241.

Farnworth, C.C. and E.A. Dratz. 1976. Oxidative damage of retinal rod outer segment membranes and the role of vitamin E. Biochimica Biophysica Acta 443:556-570.

Feeney, L. 1978. Lipofuscin and melanin of human retinal pigment epithelium. Invest. Ophthalmol. Vis. Sci. 17:583-600.

Feeney, L., J.A. Grieshaber and M.J. Hogan. 1965. Studies on human ocular pigment. In: The Structure of the Eye II, J.W. Rohen, ed., Schattauer-Verlag, Stuttgart, pp. 535-548.

Feeney-Burns, L., E.R. Berman and H. Rothman. 1980. Lipofuscin of human retinal pigment epithelium. Amer. J. Ophthalmol. 90:783-791.

Feeney-Burns, L., G.E. Eldred. 1983. The fate of the phagosome: conversion to 'age pigment' and impact in human retinal pigment epithelium. Trans. Ophthalmol. Soc. UK 103:416-421.

Feeney-Burns, L., E.S. Hilderbrand and S. Eldridge. 1984. Aging human RPE: morphometric analysis of macular, equatorial, and peripheral cells. Invest. Ophthalmol. Vis. Sci. 25:195-200.

Fliesler, S.J. and R.E. Anderson. 1983. Chemistry and metabolism of lipids in the vertebrate retina. Progr. Lipid Res. 22:79-131.

Hayes, K.C. 1974a. Retinal degeneration in monkeys induced by deficiencies of vitamin E or A. Invest. Ophthalmol. Vis. Sci. 13:499-510.

Hayes, K.C. 1974b. Pathophysiology of vitamin E deficiency in monkeys. Amer. J. Clin. Nutr. 27:1130-1140.

Hayes, K.C., J.E. Rousseau and D.M. Hegsted. 1970. Plasma tocopherol concentrations and vitamin E deficiency in dogs. J. Amer. Vet. Med. Assoc. 157:64-71.

Herrmann, R.K., W.G. Robison, Jr. and J.G. Bieri. 1984. Deficiencies of vitamins E and A in the rat: Lipofuscin accumulation in the choroid. Invest. Ophthalmol. Vis. Sci. 25:429-433.

Herrmann, R.K., W.G. Robison, Jr., J.G. Bieri and M. Spitznas. 1985. Lipofuscin accumulation in extraocular muscle of rats deficient in vitamins E and A. Graefe's Arch. Clin. Exp. Ophthalmol. 223:272-277.

Herron, W.L.,Jr. and B.W. Riegel. 1974. Production rate and removal of rod outer segment material in vitamin A deficiency. Invest. Ophthalmol. Vis. Sci. 13:46-53.

Hirosawa, K. and E. Yamada. 1976. Localization of vitamin A in the mouse retina as revealed by radioautography. In: The Structure of the Eye III, E. Yamada and S. Mishima, eds., Japanese Journal of Ophthalmology, Tokyo, pp. 165-175.

Hogan, M.J. 1972. Role of the retinal pigment epithelium in macular disease. Trans. Amer. Acad. Ophthalmol. Otolaryngol. 76:64-80

Hunt, D.F., D.T. Organisciak, H.M. Wang and R.L.C. Wu. 1984. α-tocopherol in the developing rat retina: a high pressure liquid chromatographic analysis. Current Eye Res. 3:1281-1288.

Katz, M.L., C.M. Drea, G.E. Eldred, H.H. Hess and W.G. Robison, Jr. 1986b. Influence of early photoreceptor degeneration on lipofuscin in the retinal pigment epithelium. Exp. Eye Res. 43:(in press).

Katz, M.L., C.M. Drea and W.G. Robison, Jr. 1986c. Relationship between dietary retinol and lipofuscin in the retinal pigment epithelium. Mechanisms Ageing and Development .

Katz, M.L., K.R. Parker, G.J. Handelman, T.L. Bramel and E.A. Dratz. 1982. Effects of antioxidant nutrient deficiency on the retina and retinal pigment epithelium of albino rats: a light and electron microscopic study. Exp. Eye Res. 34:339-369.

Katz, M.L. and W.G. Robison, Jr. 1984. Age-related changes in the retinal pigment epithelium of pigmented rats. Exp. Eye Res. 38:137-151.

Katz, M.L. and W.G. Robison, Jr. 1986. Nutritional influences on autoxidation, lipofuscin accumulation, and aging. In: Free Radicals, Aging, and Degenerative Diseases, J.E. Johnson, Jr., R. Walford. D. Harman and J. Miquel, eds., Alan Liss, Inc., New York, pp. 221-259.

Katz, M.L., W.G. Robison, Jr. and E.A. Dratz. 1984a. Potential role of autoxidation in age changes of the retina and retinal pigment epithelium of the eye. In: Free Radicals in Molecular Biology, Aging, and Disease, D. Armstrong, R.S. Sohal, R.G. Cutler and T.S. Slater, eds., Raven Press, New York, pp. 163-180.

Katz, M.L., W.G. Robison, Jr., and C.M. Drea. 1986a. Factors influencing lipofuscin accumulation in the retinal pigment epithelium of the eye. In: Age Pigment: Biological Markers in Aging and Environmental Stress, E.A. Totaro, D. Armstrong, P. Glees and F.A. Pisanti, eds., NVU Science Press, Amsterdam (in press).

Katz, M.L., W.G. Robison, Jr., R.K. Herrmann, A.B. Groome and J.G. Bieri. 1984b. Lipofuscin accumulation resulting from senescence and vitamin E deficiency: spectral properties and tissue distribution. Mech. Ageing Develop. 25:140-159.

Katz, M.L., W.L. Stone and E.A. Dratz. 1978. Fluorescent pigment accumulation in retinal pigment epithelium of antioxidant-deficient rats. Invest. Ophthalmol. Vis. Sci. 17:1049-1058.

Kolb, H. and P. Gouras. 1974. Electron microcopic observations of human retinitis pigmentosa, dominantly inherited. Invest. Ophthalmol. vis. Sci. 13:487-498.

LaVail, M.M. 1976. Rod outer segment disk shedding in rat retina: relationship to cyclic lighting. Science 194:1071-1074.

LaVail, M.M. 1979. The retinal pigment epithelium in mice and rats with inherited retinal degeneration. In: The Retinal Pigment Epithelium, K.M. Zinn and M.F. Marmor, eds., Harvard University Press, Cambridge, Mass., pp. 357-380.

LaVail, M.M., R.L. Sidman and D. O'Neil. 1972. Photoreceptor-pigment epithelial cell relationships in rats with inherited retinal degeneration. J. Cell Biol. 53:185-209.

Malatesta, C. 1951. Le alterazioni istopatologiche dell'occhio nella avitaminosi "E" sperimentale dei ratti ed in altre condizioni di dieta carenziale sintetica. Bolletino d'Oculistica 30:541-552.

Mann, D.M.A. and P.O. Yates. 1974. Lipoprotein pigments—their relationship to ageing in the human nervous system, I. The lipofuscin content of nerve cells. Brain 97:481-488.

Marmor, M.F. 1979. Dystrophies of the retinal pigment elpithelium. In: The Retinal Pigment Epithelium, K.M. Zinn and M.F. Marmor, eds., Harvard University Press, Cambridge, Mass., pp. 424-453.

Miquel, J., J. Oro, K.G. Bensch and J.E. Johnson, Jr. 1977. Lipofuscin: fine-structural and biochemical studies. In: Free Radicals in Biology, Volume III, W.A. Pryor, ed., Academic Press, New York, pp. 133-182.

Mullen, R.J. and M.M. LaVail. 1976. Inherited retinal dystrophy: Primary defect in pigment epithelium determined with experimental rat chimeras. Science 192:799-801.

Neville, H., D. Armstrong, B. Wilson, N. Koppang and C. Wehling. 1980. Studies on the the retina and the pigment epithelium in hereditary canine ceroid lipofuscinosis III. Morphologic abnormalities in retinal neurons and retinal pigmented epithelial cells. Invest. Ophthalmol. Vis. Sci. 19:75-86.

Noell, W.K., M.C. Delmelle and R. Albrecht. 1971. Vitamin A deficiency effect on retina: dependence on light. Science 172:72-80.

Pearse, A.G. 1972. Histochemistry, Theoretical and Applied, Third Edition, volume 2. Churchill Livingstone, Edinburgh pp. 1076-1100.

Popper, H. 1944. Distribution of vitamin A in tissue as visualized by fluorescence microscopy. Physiol. Rev. 24:205-224.

Porta, E.A. and W.S. Hartroft. 1969. Lipid pigments in relation to aging and dietary factors (lipofuscins). In: Pigments in Pathology, M. Wolman, ed., Academic Press, New York, pp. 191-235.

Reme, C.E. 1977. Autophagy in visual cells and pigment epithelium. Invest. Ophthalmol. Vis. Sci. 16:807-814.

Riis, R.C., B.E. Sheffy, E. Loew, T.J. Kern and J.S. Smith. 1981. Vitamin E deficiency retinopathy in dogs. Amer. J. Vet. Res. 42:74-86.

Robison, W.G., Jr. and T. Kuwabara. 1977. Vitamin A storage and peroxisomes in retinal pigment epithelium and liver. Invest. Ophthalmol. Vis. Sci. 16:1110-1117.

Robison, W.G., Jr., T. Kuwabara and J.G. Bieri. 1979. Vitamin E deficiency and pigment epithelial changes. Invest. Ophthalmol. Vis. Sci. 18:683-690.

Robison, W.G., Jr., T. Kuwabara and J.G. Bieri. 1980. Deficiencies of vitamins E and A in the rat, retinal damage and lipofuscin accumulation. Invest. Ophthalmol. Vis. Sci. 19:1030-1037.

Robison, W.G., Jr., T. Kuwabara and J.G. Bieri. 1982. The roles of vitamin E and unsaturated fatty acids in the visual process. Retina 2:263-281.

Rosenkranz, J. 1977. New aspects of the ultrastructure of frog rod outer segment. Inter. Rev. Cytology 50:25-158.

Schairer, E. and K. Patzelt. 1940. [Studies on the vitamin A metabolism of the eye in laboratory animals using the luminescence microscope]. Virchows Arch (Pathol. Anat.) 307:124-150.

Siakotos, A.N. and K.D. Munkres. 1981. Purification and properties of age pigments. In: Age Pigments, R.S. Sohal, ed., Elsevier/North-Holland Biomedical Press, Amsterdam, pp. 181-202.

Sickle, W. 1972. Retinal metabolism in dark and light. In: Handbook of Sensory Physiology, Volume VII, part 2, M.G.F. Fuortes, ed., Springer-Verlag, Berlin, pp. 667-727.

Sohal, R.S., ed. 1981. Age Pigments. Elsevier/North-Holland Biomedical Press, Amsterdam.

Stone, W.L., C.C. Farnsworth and E.A. Dratz. 1979. A reinvestigation of the fatty acid content of bovine, rat and frog retinal rod outer segments. Exp. Eye Res. 28:387-397.

Szamier, R.B., E.L. Berson, R. Klein and S. Meyers. 1979. Sex-linked retinitis pigmentosa: ultrastructure of photoreceptors and pigment epithelium. Invest. Ophthalmol. Vis. Sci. 18:145-160.

Tappel, A.L. 1975. Lipid peroxidation and fluorescent molecular damage to membranes. In: Pathobiology of Cell Membranes, B.F. Trump and A.U. Arstila, eds., Academic Press, New York, pp. 145-172.

Tappel, A.L., B. Fletcher and D. Deamer. 1973. Effect of antioxidants and nutrients on lipid peroxidation fluorescent products and aging parameters in the mouse. J. Gerontol. 28:415-424.

Wald, G. 1968. Molecular basis of visual excitation. Science 162:230-239.

Wiegand, R.D., N.M. Giusto, L.M. Rapp and R.E. Anderson. 1983. Evidence for rod outer segment lipid peroxidation following constant illumination of the rat retina. Invest. Ophthalmol. Vis. Sci. 24:1433-1435.

Wiggert, B., E. Masterson, P. Israel and G.J. Chader. 1979. Differential retinoid binding in chick pigment epithelium and choroid. Invest. Ophthalmol. Vis. Sci. 18:306-310.

Wing, G.L., G.C. Blanchard and J.J. Weiter. 1978. The topography and age relationship of lipofuscin concentration in the retinal pigment epithelium. Invest. Ophthalmol. Vis. Sci. 17:601-607.

Wolfe, L.S., N.M.K.N.Y. Kin, R.R. Baker, S. Carpenter and F. Andermann. 1977. Identification of retinoyl complexes as the autofluorescent component of the neuronal storage material in Batten disease. Science 195:1360-1362.

Young, R.W. 1971. The renewal of rod and cone outer segments in the rhesus monkey. J. Cell Biol. 49:303-318.

Young, R.W. and D. Bok. 1969. Participation of the retinal pigment epithelium in the rod outer segment renewal process. J. Cell Biol. 42:392-403.

Young, R.W. and D. Bok. 1979. Metabolism of the retinal pigment epithelium. In: The Retinal Pigment Epithelium, K.M. Zinn and M.F. Marmor, eds., Harvard University Press, Cambridge, Mass., pp. 103-123.

Zeman, W. 1971. The neuronal ceroid-lipofuscinoses-Batten-Vogt syndrome: a model for human aging? Adv. Gerontol. Res. 3:147-170.

Zigler, J.S. and H.H. Hess. 1985. Cataracts in the Royal College of Surgeons rat: evidence for initiation by lipid peroxidation products. Exp. Eye Res. 41:67-76.

Zinn, K.M. and M.F. Marmor, eds. 1979. The Retinal Pigment Epithelium. Harvard University Press, Cambridge, pp. 521.

Zuckerman, R. and J.J. Weiter. 1980. Oxygen transport in the bullfrog retina. Exp. Eye Res. 30:117-127.

Reactive Oxygen Species-Toxicity, Metabolism, and Reactions in the Eye

Barbara Buckley and Bruce Freeman

Oxygen, while providing the basis for aerobic life, is the source of partially reduced species which can react with biological molecules and lead to deleterious consequences for cells and tissues. Thus oxygen, due to its ubiquitous nature and its reactive intermediates superoxide (O_2^-) hydrogen peroxide (H_2O_2), hydroxyl radical (OH·), and singlet oxygen ($^1O_2$), can be toxic. Cells and tissues cope with reactive oxygen through an interconnected system of enzymatic and non-enzymatic defenses. Mechanisms of injury caused by these reactive intermediates, and biological defenses against them, will be reviewed in this chapter with special attention paid to the eye.

FORMATION OF REACTIVE OXYGEN SPECIES

While molecular oxygen (O_2) has strong oxidizing potential, direct reaction of O_2 with most biological molecules is not permitted due to spin restrictions on its electrons (Green and Hill, 1984). During normal mitochondrial respiration, the reduction of O_2 to H_2O is catalyzed by cytochrome c oxidase and involves the direct transfer of four electrons (Chance et al., 1979). Greater than 85% (Freeman and Crapo, 1981) of biological O_2 is consumed in this manner without the release of free radical intermediates. However, partial reduction of oxygen can occur by enzymatic catalysis, during autoxidation of small macromolecules and by components of electron transport systems (Freeman and Crapo, 1982). This can result in the stepwise production of reactive oxygen species in virtually every cell compartment.

$$O_2 \xrightarrow{1e^-} O_2^{\cdot -} \xrightarrow[2H^+]{1e^-} H_2O_2 \xrightarrow{1e^-} OH^- + OH^\cdot \xrightarrow[2H^+]{1e^-} 2\,H_2O \qquad \text{Equation 1}$$

H_2O_2, although chemically not a radical species because it contains no unpaired electrons, is a stronger oxidant than O_2 and shares with $O_2^{\cdot -}$ and $OH^\cdot$ the ability to react with other molecules and produce radicals. (Pryor, 1985).

Enzymatic sources of $O_2^{\cdot -}$ include xanthine oxidase (Fridovich, 1970), aldehyde oxidase (Rajagopalan, 1980), flavin dehydrogenases (Massey et al., 1969) and dihydroorotate dehydrogenase (Aleman and Handler, 1967). In activated phagocytic cells, plasma membrane-associated NADPH oxidase generates and releases $O_2^{\cdot -}$ into the extracellular milieu, thereby providing a reactive oxygen component of inflammation (Babior and Peters, 1981). Superoxide can also be produced in cells by autoxidation of numerous small molecules such as hemoglobin (Misra and Fridovich, 1972), myoglobin (Gotoh and Shikama, 1976), reduced ferredoxins (Misra and Fridovich, 1971), flavins (Ballou et al., 1969), tetrahydropterins (Nishimiki , 1975), cateocholamines (Misra and Fricovich, 1972), polyhydric phenols (Marklund, 1984) and thiols (Baccaneri, 1978). Some of these oxidation-reduction reactions are catalyzed by transition metals (Bacaneri, 1978; Misra and Fridovich, 1972). Components of electron transport chains, present in mitochondria and endoplasmic reticulum, can also serve as sources of $O_2^{\cdot -}$. In the mitochondrial electron transport system, both the NADH dehydrogenase site and the ubiquinone-cytochrome b site are capable of reducing of O_2 to form $O_2^{\cdot -}$ (Turrens et al., 1982; Cadenas et al., 1977). Mitochondrial production of $O_2^{\cdot -}$ is increased during hyperoxic conditions or when electron transport component are highly reduced (Turrens et al., 1982). The latter case occurs in the presence of electron transport inhibitors or when the availability of ADP and O_2 is limited, leading to the accumulation of reduced electron transport components. Superoxide radical can also be generated by electron transport chains of endoplasmic reticulum and nuclear membrane (which is continuous with the endoplasmic reticulum) including cytochromes P450, b_5 and their associated reductases (Estabrook and Werringloer, 1976). These enzymes hydroxylate nonpolar compounds via electron transfer from NADPH and NADP. Certain substrates for these enzymes are capable of diverting electrons from associated electron transport carriers and undergoing redox-cycling, resulting in the production of O_2^- (Pryor, 1985). Such substrates include quinoid and anthracycline antibiotics, bleomycin (Doroshow and Hochstein, 1982) and paraquat (Mason, 1982).

The production of H_2O_2 within cells is primarily due to the dismutation of O_2^- (Equation 2) and divalent reduction of O_2 to H_2O_2 by peroxisomal enzymes. Superoxide dismutation can occur spontaneously or be catalyzed by the enzyme superoxide dismutase (Fridovich, 1983):

$$O_2^- + O_2^- \xrightarrow{\ 2H^+\ } H_2O_2 + O_2 \qquad\qquad \text{Equation 2}$$

Consequently, O_2^- and its metabolite H_2O_2 are often co-distributed within cells and cellular compartments. Specialized organelles called peroxisomes contain a number of oxidases which catalyze the divalent reduction of O_2 to form H_2O_2 during substrate oxidation. D-amino acid oxidase, 1-alpha-hydroxyacid oxidase and fatty acyl-coA oxidase are all specific generators of H_2O_2 (Masters and Holmes, 1977).

Hydroxyl radical (OH·) can be produced by the pseudo Haber-Weiss reaction (Equations 3-5) involving O_2^- and H_2O_2 in the presence of a metal ion or by Fenton's reaction (Equation 5) involving H_2O_2 and a metal ion (Aust and Thomas, 1985).

$$O_2^- + Fe^{3+} \longrightarrow O_2 + Fe^{2+} \qquad\qquad \text{Equation 3}$$

$$O_2^- + O_2^- \xrightarrow{\ +2H^+\ } O_2 + H_2O_2 \qquad\qquad \text{Equation 4}$$

$$Fe^{2+} + H_2O_2 \longrightarrow Fe^{3+} + OH· + OH- \qquad\qquad \text{Equation 5}$$

Transition metals, like iron and copper, serve as good catalysts of oxidation-reduction reactions by acting as intermediates in electron transfer. Some chelators, like EDTA, promote these reactions by enhancing the solubility of the metal without tying up all the available coordination sites of the metal (Aust and Thomas, 1985). Hydroxyl radical can also be produced by irradiation of tissues, which results in the homolytic cleavage of water (Bielski and Gebieki, 1977).

Another reactive oxygen species is singlet oxygen $^1O_2$. Singlet oxygen exists at the same oxidation level as O_2, but has a higher state of electronic excitation. Whether enzymatic sources of $^1O_2$ exist in eukaryotic systems is a subject of controversy. However, $^1O_2$ is known to be produced in photochemical reactions involving the interaction of photons with chromophores and O_2 (Foote, 1976). At this point, the biological significance of $^1O_2$ in eukaryotes is not well established.

REACTIONS OF REDUCED OXYGEN SPECIES

Factors influencing the relative toxicity of O_2^-, H_2O_2, OH· and $^1O_2$ include their reactivity with biomolecules, stability and ability to diffuse through various cell compartments. Singlet O_2 and OH· are chemically the most reactive species and are characterized by the shortest half-lives (Pryor, 1985). Since neither of these reactive species will diffuse more than a few times their molecular radii before reacting, interaction of OH· and $^1O_2$ with critical biological targets is likely to be dependent on the promixity of the generation site to the target molecule. Hydrogen peroxide, on the other hand, is capable of diffusing relatively long distances due to its stability (Pryor, 1985). Since biological membranes are permeable to H_2O_2, this species can diffuse between various cellular compartments (deDuve and Baudhuin, 1965; Nicholls, 1965). It has been demonstrated that H_2O_2 can diffuse from peroxisomes and mitochondria to cytosol and escape from the cytosol to extracellular spaces (Jones, 1982). Although membranes are relatively impermeable to superoxide anion, the protonated form of O_2^- may cross membranes (Fridovich, 1983). This species, the perhydroxyl radical (HO_2·), is possibly a stronger oxidant than the nonprotonated O_2^-. Transmembrane O_2^- movement may additionally be facilitated by anion channels, as demonstrated in erythrocytes (Lynch, 1978). In summary, H_2O_2 and O_2^- may affect target molecules at some distance from their generation sites. The differences in stability, reactivity and diffusibility of reduced oxygen species can often make it difficult to determine which species is the most toxic or which is responsible for a given toxic effect.

Reactive oxygen species can cause damage to cellular components through a number of different mechanisms. Superoxide and its metabolites can alter the function of cellular macromolecules through oxidizing chain reactions, as in the case of oxyhemoglobin (Lynch and Fridovich, 1976) and NADH which is bound to lactate dehydrogenase (Bielski and Gebieki, 1974). Proton abstraction by partially reduced oxygen results in the formation of carbon and sulfur-centered radicals as shown in Equation 6 (Pryor, 1985).

$$XH + O_2^{\cdot-} \longrightarrow X^{\cdot-} + O_2 + H^+ \qquad \text{Equation 6}$$

Unsaturated bonds of fatty acids as well as aromatic and thiol-containing amino acids are particularly susceptible to oxidation (Freeman and Crapo, 1982). Oxidation of amino acids or hydroxylation of amino acid side chains can affect enzymatic activity and protein function. In membrane phospholipids, the ratio of unsaturated to saturated fatty acids is a key factor in determining the susceptibility of the membrane membrane to peroxidative processes.

Peroxidation of membrane phospholipids involves a series of often metal catalyzed chain reactions and is thought to be initiated by $O_2^{\cdot-}$ or a metaldioxygen complex (Aust and Thomas, 1985). In the first step (Equation 7), the reactive oxygen species abstracts a proton from the allylic position of an unsaturated bond. Diene conjugation occurs, followed by the incorporation of O_2 and the formation of a peroxyl radical ($ROO^{\cdot}$, Equation 8). The peroxyl radical can further react by abstracting hydrogen ions from adjacent unsaturated fatty acids (Equation 9).

$$OH^{\cdot} + RH \longrightarrow H_2O + R^{\cdot} \qquad \text{Equation 7}$$

$$R^{\cdot} + O_2 \longrightarrow ROO^{\cdot} \qquad \text{Equation 8}$$

$$ROO^{\cdot} + R\,H \longrightarrow ROOH + R^{\cdot} \qquad \text{Equation 9}$$

The lipid hydroperoxides (ROOH) formed, like H_2O_2, can potentially decompose in the presence of transition metals (Equations 10 and 11), again forming radical species (Aust and Thomas, 1985).

$$ROOH + Fe^{2+} \longrightarrow RO^{\cdot} + OH^- + Fe^{3+} \qquad \text{Equation 10}$$

$$Fe^{3+} + ROOH \longrightarrow ROO^{\cdot} + H^+ + Fe^{2+} \qquad \text{Equation 11}$$

These cyclical reactions ultimately lead to the oxidative degradation of phospholipids. Lipid peroxidation introduces hydrophilic components in the form of lipid hydroperoxides and their lipid alcohol metabolites into the membrane, thus altering membrane fluidity (Chance et al., 1979). The production of hydroperoxides can also affect the activity of peroxide-sensitive enzymes such as cyclooxygenase (Hemler et al., 1979). Furthermore, aldehydes formed as by-products of lipid peroxidation (e.g. malondialdehyde) can react with primary amines of nearby lipids or proteins, forming conjugated Schiff's base products and cross-linking membrane components. The "age pigment" lipofuscin is believed to partly represent accumulation of these conjugated Schiff's base products in storage depots such as lysosomes (Chance et al., 1979). Through these reactions with proteins and lipids, oxygen metabolites can affect not only membrane fluidity and permeability, but also membrane-associated transport and receptor functions.

Reactive oxygen species can also react with polysaccharides and sugars causing a variety of effects. Decrements in cell surface function can occur following oxidation of transmembrane glycoproteins. Reaction of oxidants with extracellular matrix proteoglycans and polymers (e.g. collagen and hyaluronate) can result in alteration of physical state (Greenwald and Moy, 1980; Riley and Kerr, 1985). In addition, DNA strand scissions can occur due to the oxidation of sugars which form the backbone of nucleotides (Brawn and Fridovich, 1981).

DEFENSES AGAINST REACTIVE OXYGEN SPECIES

Given the diverse mechanisms by which reactive oxygen species can exert deleterious effects on biological target molecules, effective tissue defenses against oxidant reactions are essential for viability and maintenance of cellular homeostasis. One set of defenses lies in the control of oxygen delivery to cells and tissues. Except for certain regions of lung and eye, tissue and cellular oxygen tensions are controlled by the microvascular system and by characteristics of oxygen-hemoglobin binding (Chance et al., 1979).

Three enzymes are primarily involved in maintaining low levels of O_2^- and H_2O_2 in cells: superoxide dismutase, catalase and glutathione peroxidase. Superoxide dismutase (SOD) catalyzes the dismutation of O_2^- and H_2O_2 as a pseudo-first order process. With a rate constant of 2×10^9 M^{-1} sec^{-1} at pH 7.8, superoxide dismutase operates at near diffusion-controlled rates (Fridovich, 1983). Spontaneous dismutation occurs as a second order process with a rate constant of 8×10^4 $M^{-1} \cdot sec^{-1}$ at pH 7.8 (Fridovich, 1983). Nonenzymatic dismutation is highly pH dependent and is more likely to occur under conditions which favor the protonation of O_2^- (pK_a of 4.8), since electrostatic repulsion makes interaction of two superoxide anions unlikely. Enzymatic dismutation, on the other hand, is facilitated by the close proximity of O_2^- and ligated metal ions at the enzyme active site (Fridovich, 1983).

Superoxide dismutase exists as three metalloprotein isozymes having either copper-zinc, manganese or iron at the active site (Fridovich, 1983). These metals perform catalytic functions by facilitating electron transfer during dismutation (Equations 12 and 13.)

$$E\text{-}Cu^{2+} + O_2^- \longrightarrow E\text{-}Cu^{1+} + O_2 \qquad \text{Equation 12}$$

$$E\text{-}Cu^{1+} + O^- \longrightarrow E\text{-}Cu^{2+} + H_2O_2 \qquad \text{Equation 13}$$

While iron and manganese-containing superoxide dismutases are usually found in prokaryotes, the copper/zinc and manganese forms are characteristic of eukaryotic systems (Weisiger and Fridovich, 1973). The latter two metalloenzymes usually follow different distributions in eukaryotic cells, with manganese superoxide dismutase primarily localized in mitochondria and copper-zinc superoxide dismutase primarily localized in cytosol (Fridovich, 1983). An extracellular copper and possibly zinc-containing form of the enzyme has also been reported (Marklund, 1984).

The metabolism of H_2O_2 to H_2O plus O_2 is catalyzed primarily by catalase and glutathione peroxidase. Catalase, which contains heme-iron at its active site to facilitate electron transfer, is localized in the cytosol of erythrocytes and in peroxisomes of other cells (Chance et al., 1979). It can operate in two modes, peroxidatic (alcohol metabolism) and catalatic (H_2O_2 metabolism). Due to its kinetic mechanism, the peroxidatic mode predominates when levels of H_2O_2 are low and alcohol is available, while the catalatic activity mode predominates when levels of H_2O_2 are high (Chance et al., 1979). For this reason, catalase is especially important as a cellular defense when intracellular generation of H_2O_2 is high.

Studies of prokaryotes have demonstrated the evolutionary significance of superoxide dismutase and catalase. Most oxygen-tolerant prokaryotes contain superoxide dismutase and catalase, while very low levels of these enzymes are found in anaerobic microorganisms (Fridovich, 1983; Chance, 1979). Synthesis of superoxide dismutase and catalase can be induced in prokaryotes by increasing the partial pressure of O_2 under which the organisms are grown or by increasing the intracellular production of O_2^- at constant partial pressures of O_2. Increased activities of superoxide dismutase and catalase have also been observed in tissues of eukaryotic organisms exposed to increased partial pressure of O_2 and other conditions presumed to increase the intracellular level of O_2^- (Kimball et al., 1976; Crapo and Tierney, 1974; Frank, 1982). These enzymatic elevations observed in both prokaryotes and eukaryotes represent adaptive responses to increased concentrations of reactive oxygen species.

Glutathione peroxidase, which exists in selenium-containing and non-selenium-containing forms, also metabolizes H_2O_2 and other hydroperoxides. The subcellular distribution of the major selenium-containing glutathione peroxidase differs from catalase in that it is localized in both cytosol (70%) and the mitochondrial matrix (30%) (Wendel, 1980). Although its rate constant of 5×10^7 $M^{-1}sec^{-1}$ is similar to that of catalase (Flohe et al., 1972; Chance et al., 1979), it has a lower Km for H_2O_2 than catalase (Cohen and Hochstein, 1963). Hence glutathione peroxidase is the predominant pathway for H_2O_2 metabolism at low concentrations of substrate. Glutathione peroxidase has an absolute requirement for reduced glutathione (GSH) as a cofactor, but is non-specific for its peroxide substate, metabolizing H_2O_2 as well as alkyl hydroperoxides (Chance et al., 1979).

These three enzymatic defenses also protect each other against cross-inactivation by their subtrates. Superoxide dismutase defends catalase and glutathione peroxidase from inactivation by O_2^-, (Blum and Fridovich, 1984; Kono and Fridovich, 1982; Rister and Baehner, 1976) while catalase defends superoxide dismutase from inactivation by H_2O_2 (Bray et al., 1974; Hodgson and Fridovich, 1975).

Glutathione peroxidase plays an important role in the metabolism of hydroperoxide species which are derived from lipid peroxidation chain reactions (Chance et al., 1979). A membrane-associated glutathione peroxidase activity specific for lipid hydroperoxides has also recently been reported (Ursini et al., 1985). Propagation of free radical reactions by the decomposition of hydroperoxides to hydroperoxyl radicals can be prevented by the catalytic action of glutathione peroxidase. This reaction requires GSH as a source of reducing equivalents and results in its oxidation (Equation 14). Glutathione disulfide (GSSG) is subsequently reduced by glutathione reductase (GR) using NADPH as a source of reducing equivalents (Equation 15).

$$\text{ROOH (H}_2\text{O}_2\text{)} + \text{2GSH} \xrightarrow{\text{GP}} \text{ROH (H}_2\text{O)} + \text{GSSG} + \text{H}_2\text{O} \qquad \text{Equation 14}$$

$$\text{GSSG} + \text{NADPH} + \text{H}^+ \xrightarrow{\text{GR}} \text{2GSH} + \text{NADP}^+ \qquad \text{Equation 15}$$

Glutathione represents the major source of cellular nonprotein thiol groups while the enzyme glutathione reductase is distributed in cells similarly to glutathione peroxidase (Flohe and Schlegel, 1971). The coupled glutathione peroxidase and glutathione reductase reactions, along with reducing equivalents from NADPH, are collectively termed the glutathione redox cycle. Enzymatic sources of NADPH include dehydrogenases of the pentose phosphate shunt, malic enzyme, isocitrate dehydrogenase and NADPH-NADH transhydrogenase (Chance et al., 1979). Kinetic studies have demonstrated that while the metabolism of hydroperoxides by the first order glutathione peroxidase reaction is rapid, the cycle may be rate limited by the slow reduction of GSSG catalyzed by glutathione reductase (Wendel, 1980). This may be due to insufficient NADPH and may lead to elevated intracellular levels of GSSG. Since hydroperoxide metabolism is coupled to NADPH oxidation through the glutathione redox cycle, it can be seen that a high rate of glutathione peroxidase turnover may affect other NADPH-requiring processes (e.g. lipogensis) of the cell. Additionally, NAD-requiring processes, like mitochondrial respiration, may be affected by glutathione peroxidase turnover via the NADPH-NADH transhydrogenase reaction (Chance et al., 1979).

Elevated intracellular levels of GSSG as a result of hydroperoxide metabolism can alter the thiol:disulfide status of proteins. Disulfide exchange reactions between GSSG and protein thiol groups can result in protein inactivation and inhibition of cell division (Kosower and Fridovich, 1975; Barron, 1951).

$$GSSG + PSH \longrightarrow GSH + PSSG \qquad \text{Equation 16}$$

Some cells and tissues, including erythrocytes (Srivastava and Beutler, 1969a), crystallin lens (Srivastava and Beutler, 1969b), liver (Sies and Summer, 1975) and lung (Nishiki et al., 1976) protect themselves against these toxic effects of GSSG accumulation via extracellular transport mechanisms for GSSG.

Non-enzymatic defenses against oxygen radical-mediated injury also exist. In the cytosol, low molecular weight reductants such as glutathione and ascorbate serve as radical scavegers (Freeman and Crapo, 1972). In lipid soluble compartments, termination of free radical chain reactions is mediated by α-tocopherol (Burton and Ingold, 1984; Nishimiki, 1980). Ascorbate also functions to maintain α-tocopherol in the reduced state (Tappel, 1969). Beta-carotene quenches lipid free radical reactions at low partial pressures of O_2, while at high partial pressures it may act as a pro-oxidant (Burton, 1984). In addition, β-carotene serves as an efficient scavenger of reactive oxygen species (Burton, 1984).

OXYGEN-MEDIATED PHOTOCHEMICAL INJURY TO PHOTORECEPTOR CELLS

The elongate and cylindrical achromatic dim light receptor cell of the eye is termed the rod. For morphological and functional reasons, the rod may be divided into four separate parts: the outer segment, inner segment, nuclear region and synaptic pedicle. The rod outer segment is particularly sensitive to photochemical injury induced by excessive exposure to light and secondary to use of lasers in surgical procedures on the eye. When photons of light are absorbed by molecules, the molecule can exist in a transient, excited state such as observed in singlet oxygen. Molecules in a transient excited state can also transfer excess energy to adjacent molecules up to 10 nanometers apart. Rod outer segments are susceptible to oxidant injury because their membrane phospholipids contain the highest concentration of long chain polyunsaturated fatty acids of any membrane system ever studied, making them prone to peroxidation (Fliesler and Anderson, 1983).

Rod outer segments contain high specific activities of superoxide dismutase and relatively low specific activities of glutathione peroxidase and catalase, suggesting a potential to accumulate H_2O_2. As discussed earlier, hydroxyl radical derived from hydrogen peroxide is an efficient initiator of lipid peroxidation. Significant losses in docosahexaenoic acid (22:6ω3) and increased lipid peroxides in rod outer segments have been observed in light-induced retinal degeneration in albino rats confirming the concept of peroxidation susceptibility (Wiegand, 1984). Similar losses of docosahexaenoic acid in whole retinas and decreased α-tocopherol levels occurred in rabbit eyes after constant illumination, suggesting oxidant stress (Joel et al., 1981). Although causal and coincidental effects are difficult to distinguish, chemical and morphological studies suggest that oxidative processes mediated in part by lipid peroxidation are associated with retinal degeneration. The retina also has the highest rate of oxygen consumption by weight of any tissue and is exposed to a prodigious oxygen tension. The observation that increasing the arterial PO_2 in rhesus monkeys increased retinal sensitivity to photochemical light damage by as much as a factor of 3 (Ham et al., 1984) is also presumptive evidence that reactive oxygen species are contributory mediators of retinal photochemical injury.

CATARACTS AND OXIDANT-INDUCED LENTICULAR DEGENERATION

Oxidant damage to lenticular protein is recognized as a major event in cataractogenesis. There is increased oxidation of lens protein amino acids and formation of protein-glutathione mixed disulfides and protein-protein disulfides which is correlated with both age of an individual and progression of cataracts. The generation of high molecular weight protein aggregates results in light scattering. In senile, diabetic and certain expermentally-induced cataracts, protein aggregation is accompanied by other post-translational changes of lens proteins including oxidation of methionine to methionine sulfoxide, oxidation of tryptophan to N-formyl kyneurenine and tyrosine to dihydroxyphenylalanine. These events contribute to the abnormal fluorescence and color characteristics of the cataractous lens (Garner and Spector, 1980).

Hydrogen peroxide is present in relatively high concentrations of 50 to 700 µM in the aqueous humor of various species (Garner et al., 1983). Both hydrogen peroxide and H_2O_2-derived OH· are strong candidates for inducing the protein oxidation found in cataractous lenses. The protective action of reduced glutathione and the glutathione redox cylcle would be important in preventing excessive hydrogen peroxide accumulation in lens and subsequent toxic reactions. Depletion of rat lens glutathione by incubation with 1-chloro-2,4-dinitrobenzene renders lenses more susceptible to in vitro cataractogenesis induced by oxidants including H_2O_2 and O_2^- (Ansari and Srivastava, 1982). Interestingly, glutathione can also play a protective role against cataractogenesis, following oxidation to the disulfide form. Glutathione disulfide can undergo mixed disulfide exchange with lens proteins to form protein-glutathione mixed disulfides. Formation of mixed disulfides could then retard or prevent the disadvantageous formation of lens crystallin protein-protein disulfides (Mostafapour and Reddy, 1982).

Lipid peroxidation is also associated with cataracts. Malondialdehyde, a breakdown product of lipid peroxides, is increased 4-fold in advanced cataracts (Bhuyan et al., 1982). Thus, one would expect increased membrane permeability and altered transport of low molecular weight ions in cataracts. This appears to be the case, since Na^+, K^+ and Ca^{2+} concentrations are altered in cataracts. The aqueous fluid surrounding the lenses in patients with cataracts has dramatically elevated concentrations of H_2O_2 (Garner et al., 1983). This suggests that elevated exogeneous H_2O_2 is an important oxidizing agent which can influence lipid peroxidation in lenses and the formation of certain cataracts. Support for this is provided by the observation that cultured lenses exposed to H_2O_2 concentrations similar to those found in the aqueous humor will result in development of lens opacity within two days (Spector and Garner, 1982). In support of this observation, Mg^{2+}-stimulated Na^+-ATPase is uncoupled from ATP hydrolysis following a one hour exposure of cultured bovine lenses to 1 mM H_2O_2 (Garner et al., 1983).

Selective inhibition of enzymes which metabolize reactive oxygen species, specifically superoxide dismutase, catalase and glutathione peroxidase, has been observed in advanced cataract (Bhuyan et al., 1982). Thus, overproduction of reactive oxygen species during cataractogenesis may lead to inhibition of these oxidant inhibitable antioxidant enzymes.

OXYGEN-ASSOCIATED RETINOPATHY OF PREMATURITY

Retrolental fibroplasia is often referred to as retinopathy of prematurity. Oxygen therapy of low birth weight neonates is etiologically related to retinopathy of prematurity. Of the 20,000 neoates weighing less than 1,500 grams at birth who survive each year in the United States, 75% develop some degree of retinopathy of prematurity. In the majority, this resolves spontaneously, but about 2,000 of these infants develop permanent ocular damage and approximately 500 become legally blind.

The onset of oxygen-associated retinopathy consists of two stages. The first stage occurs during prolonged breathing of elevated oxygen concentrations and involves closure of the immature retinal vasculature. Vascular growth from the optic disc to the retinal periphery is halted during vascular closure. In the second stage, when infants are returned to ambient air, retinal vessels not permanently occluded undergo neovascular proliferation. These new vessels lack normal structural integrity and are prone to hemorrhage.

It is unclear whether the capilary loss in the oxygen treated premature infant is due to destruction of endothelial tissue, cytotoxic effects of reactive oxygen species produced during hyperoxia, or if vascular degeneration is secondary to autoregulatory effects caused by vasoconstriction and diminished blood flow. It is possible that prostaglandins or reactive oxygen species are involved in the autoregulatory changes occurring in retinopathy of prematurity (Flower and Blake, 1981). Recent clinical trials have shown that the free radical scavenger α-tocopherol has significant therapeutic efficacy in preventing retinopathy of prematurity (Hittner et al., 1983,1984). Until now, there has been no safe and effective means of counteracting retinopathy of prematurity due to oxygen supplementation. Other antioxidant agents and techniques for targeting free radical scavengers to the eye may play an important role in pharmacologic modification of oxidant-induced ocular injury.

136

REFERENCES

Aleman, V. and P. Handler. 1967. Dihydroorotate dehydrogenase. J. Biol. Chem. 242:4087-4096.

Ansari, N.H. and S.K. Srivastava. 1982. Role of glutathione in the prevention of cataractogenesis in rat lenses. Current Eye Res. 2:271-275.

Aust, S., Morehouse, L. and C. Thomas. 1985. Role of metals in oxygen radical reactions. J. Free Radicals Biol. Med. 1:3-26.

Babior, B. and W. Peters. 1981. The superoxide-producing enzyme of human neutrophils: further properties. J. Biol. Med. 256:2331-2323.

Baccaneri, D. 1978. Coupled oxidation of NADPH with thiols at neutral pH. Arch. Biochem. Biophys. 191:351-357.

Ballou, D., Palmer, G. and V. Massey. 1969. Direct demonstration of superoxide anion production during the oxidation of reduced flavin and of its catalytic decomposition by erythrocuprein. Biochem. Biophys. Res. Commun. 36:898-904.

Barron, E.S.G. 1951. Thiol groups of biological importance. Adv. Enzymol. 11:201-266.

Bhuyan, K.C. Bhuyan, D.K., Kuck, J.F.R. Jr., K.D. Kuck and H.L. Kern. 1982. Increased lipid peroxidation and altered membrane functions in Emory mouse cataract. Current Eye Res. 2:597-606.

Bielski, B. and J. Gebieki. 1977. Application of radiation chemistry to biology. In Free Radicals in Biology, Vol. 3, edited by W. Pryor. New York: Academic Press. p.1-48.

Bielski, B. and P. Chan. 1974. Kinetic study by pulse radiolysis of the lactate dehydrogenase-catalyzed chain oxidation of nicotinamide adenine dinucleotide by HO_2 and O_2-radicals. J. Biol. Chem. 250:318-321.

Blum, J. and I. Fridovich. 1984. Inactivation of glutathione peroxidase radicals. Arch. Biochem. Biophys. 240:500-508.

Brawn, K., and I., Fridovich. 1981. DNA strand scission by enzymemically generated oxygen radicals and Biochem. Biophys. 206:414-419.

Bray, R., Cockle, S., Fielden, E., Roberts, P., Rotilio, B. and L. Calabrese. 1974. Reduction and inactivation of superoxide dismutase by hydrogen peroxide. Biochem. J. 139:43-48.

Burton, G. and K. Ingold. 1984. Beta carotene: an unusual type of lipid antioxidant. Science 224:569-573.

Cadenas, E., Boveis, A., Ragan, I. and A. Stoppani. 1977. Production of superoxide radicals and hydrogen peroxide by NADH-ubiquinone reductase and ubuquinol-cytochrome c reductase from beef heart mitochondria. Arch. Biochem. Biophys. 180:248-257.

Chance, B., Sies, H., and A. Boveris. 1979. Hydroperoxide metabolism in mammalian organs. Physiol. Rev. 59:527-605.

Cohen, G. and P. Hochstein. 1963. Glutathione peroxidase: the primary agent for the elimination of H_2O_2 in erythrocytes. Biochemistry 2: 1420-1428.

Crapo, J. and D. Tierney. 1974. Superoxide dismutase and pulmonary oxygen toxicity. Am. J. Physiol. 226:1401-1407.

DeDuve, C. and P. Baudhuin. 1966. Peroxisomes (microbodies and related particles). Physiol. Rev. 323-357.

Doroshow, J. and P. Hochstein. 1982. Redox cycling and the mechanism of action of antibiotics in neoplastic diseases. In Pathology of Oxygen, edited by A. Autor. New York: Academic Press. p. 245-259.

Estabrook, R. W. and J. Werringloer. 1976. Cytochrome P_{450}: its role in oxygen activation for drug metabolism. In Drug Metabolism Concepts, edited by D. Jerna. Washington D.C.: American Chemical Society. p.1-16.

Fliesler, S. and R. Anderson. 1983. Chemistry and metabolism of lipids in the vertebrate retina. Prog. Lipid Res. 22:79-131.

Flohe, L. and W. Schlegel. 1971. Glutathione peroxidase IV. Hoppe Seylers Z. Physiol. Chem. 352:1401-1410.

Flohe, L., Loschen, G., Gunzler, W. and E. Eichele. 1972. Glutathione peroxidase V. The kinetic mechanism. Hoppe Seylers Z. Physiol. Chem. 353:987-999.

Flower, R. and D. Blake. 1981. Retrolental Fibroplasia: evidence for a role of the prostaglandin cascade in the pathogenesis of oxygen-induced retinopathy in the newborn beagle. Pediatr. Res. 15:1293-1302.

Foote, C. 1976. Photosensitized oxidation and singlet oxygen: consequences in biological systems. In Free Radicals in Biology, Vol. 2, edited by W. Pryor. New York: Academic Press. p. 85-125.

Frank, L. 1982. Pretection for O_2 toxicity by preexposure to hypoxia: lung antioxidant enzyme role. J. Appl. Physiol. 53:475-482.

Freeman, B. and J. Crapo. 1981. Hyperoxia increases oxygen radical production in rat lungs and lung mitochondria. J. Biol. Chem. 256:10986-10992.

Freeman, B. and J. Crapo. 1982. Biology of Disease: Free radicals and tissue injury. Lab. Invest. 47:412-426.

Fridovich, I. 1970. Quantitative aspects of the production of superoxide anion radical by milk xanthine oxidase. J. Biol. Chem. 245:4035-4057.

Fridovich, I. 1983. Superoxide radical: an endogenous toxicant. Ann. Rev. Pharmacol. Toxicol. 23:239-257.

Garner, M. and A. Spector. 1980. Selective oxidation of cysteine and methionine in normal and senile cataractous lenses. Proc. Natl. Acad. Sci. 77:1274-1277.

Garner, W., Garner, M. and A. Spector. 1983. H_2O_2-induced uncoupling of bovine lens Na^-, K^+-ATPase. Proc. Natl. Acad. Sci. 80:2044-2048.

Gotoh, T. and K. Shikama. 1976. Generation of the superoxide radical during the autooxidation of oxymyoglobin. J. Biochem. 80:397-399.

Green, H.J. and H.A.O. Hill (1984) Chemistry of dioxygen. Meth Enzymol. 105:3-21.

Greenwald, R. and W. Moy. Effect of oxygen-derived free radicals on hyaluronic acid. Arthritis Rheum. 23:455-463.

Greenwald, R., Moy, W. and D. Lazarus. 1976. Degradation of cartilage proteoglycans and collagen by superoxide radical. Arthritis Rheum. 19:799.

Ham, W.T., Mueller, H.A., Ruffolo, J.J., Millen, J.E., Cleary, S.F., Guerry, R.K. and Guerry, D. 1984. Basic mechanisms underlying the production of photochemical lesions in the mammalian retina. Current Eye Res. 3:165-179.

Hemler, M., Cook, H. and W. Land. 1979. Prostaglandin biosynthesis can be triggered by lipid peroxides. Arch. Biochem. Biophys. 193:340-345.

138

Hittner, H.M., Godio, L.B. Speer, M.E., Rudolph, A.J., Taylor M.M., Blifield, C., Kretzer, F.L. 1983. Retrolental fibroplasia: further clinical evidence and ultrastructural support for efficacy of vitamin E in the preterm infant. Pediatrics 71:423-32.

Hittner, H.M., Kretzer, F.L. and A.J. Rudolph. 1984. Prevention and management of retrolental fibroplasia. Hospital Practice. 19:85-94,99.

Hodgson, E. and I. Fridovich. 1975. The interaction of bovine erythrocyte superoxide dismutase with hydrogen peroxide: inactivation of the enzyme. Biochem. 14:5294-5303.

Joel. C., Briggs. S., Gaal, D., Hannan, J., Kahlow, M., Stein, M., Tarver, A. and A. Yip. 1981. Photodynamic injury to the retina of albino rabbits. Invest. Ophthalmol. Vis. Sci., ARVO Suppl., 166.

Jones, D.P. 1982. Intracellular catalase function: analysis of the catalatic activity by product formation in isolated liver cells. Arch. Biochem. Biophys. 214:806-814.

Kimball, R., Reddy, K., Pierce, T., Schwartz, L., Mustafa, M. and C. Cross. 1976. Oxygen toxicity: augmentation of antioxidant defense mechanisms in rat lung. Am. J. Physiol. 230:1425-1431.

Kono, Y. and I. Fridovich. 1982. Superoxide radical inhibits catalase. J. Biol. Chem. 256:5751-5754.

Kosower, N. and E. Kosower. 1976. Functional aspects of glutathione disulfide and hidden forms of glutathione. In Glutathione: Metabolism and Function, edited by I. Arias and W. Jakoby. New York:Raven Press. p.159-172.

Lynch, R., Lee G. and G. Cartwright. 1976. Inhibition by superoxide dismutase of methemoglobin formation from oxyhemoglobin. J. Biol. Chem. 251:1051-1019.

Lynch, R. and I. Fridovich. 1978. Effects of superoxide on the erythrocyte membrane. J. Biol. Chem. 253:1838-1845.

Marklund, S. and G. Marklund. 1974. Involvement of the superoxide anion radical in the autoxidation of pyrogallol and convenient assay for superoxide dismutase. Eur. J. Biochem. 47:469-474.

Marklund, S. 1984. Extracellular superoxide dismutase in human tissues and human cell lines. J. Clin. Invest. 74:1398-1403.

Mason, R. 1982. Free radical intermediates in the metabolism of toxic chemicals. In Free Radicals in Biology, Vol. 5, edited by W. Pryor. New York: Academic Press. p.161-222.

Massey, V., Strickland, W., Mayhew, S., Howell, L., Engel. P., Mattew, R., Schuman, M., and P. Sullivan. 1969. Direct demonstration of superoxide anion production during the oxidation of reduced flavin and of its catalytic decomposition by erythrocuprein. Biochem. Biophys. Res. Commun. 36:891-897.

Masters, C. and R. Holmes. 1977. Peroxisomes: New aspects of cell physiology and biochemistry. Physiol. Rev. 57:816-882.

Misra, H. and I. Fridovich. 1971. The generation of superoxide radical during the autoxidation of ferredoxins. J. Biol. Chem. 246:6886-6890.

Misra, H. and I. Fridovich. 1972. The role of superoxide anion in the autoxidation of epinephrine and a simple assay for superoxide dismutase. J. Biol. Chem. 247:3170-3175.

Misra, H. and I. Fridovich. 1972. The generation of superoxide radical during the autoxidation of hemoglobin. J. Biol. Chem. 247:6960-6962.

Mostafapour, M.K. and V.N. Reddy. 1982. Interactions of glutathione disulfide with lens crystallins. Current Eye Res. 2:591-596.

Nicholls, P. 1965. Activity of catalase in the red cell. Biochim. Biophys. Acta 99:286-297.

Nishiki, K., Jamieson, D., Oshino, N., and B. Chance. 1976. Oxygen toxicity in the perfused rat liver and lung under hyperbaric conditions. Biochem. J. 160:343-355.

Nishimiki, M. 1975. Generation of superoxide anion in the reaction of tetrahydropterins with molecular oxygen. Arch. Biochem. Biophs. 166:273-279.

Nishimiki, N., Yamada, H., and K. Yagi. 1980. Oxidation by superoxide by tocopherols dispersed in aqueas media with deoxycholate. Biochim. Biophys. Acta. 627:101-108.

Pryor, W. 1985. Oxy-radicals and related species: their formation, their lifetimes and their reactions. Annu. Rev. Physiol. 48:1-43.

Rajagopalan, K. 1980. Xanthine oxidase and aldehyde oxidase. In Enzymatic Basis of Detoxification, Vol. 1, edited by W. Jakoby. New York: Academic Press. p. 295-309.

Riley, D. and J. Kerr. 1985. Oxidant injury of the extracellular matrix: potential role in the pathogenesis of pulmonary emphysema. Lung 163:1-13.

Rister, M. and R. Baehner. 1976. The alteration of superoxide dismutase, catalase, glutathione peroxidase and NAD(P)H cytochrome c reductase in guinea pig polymorphonuclear leukocytes and alveolar macrophages during hyperoxia. J. Clin. Invest. 58:1174-1184.

Sies, H. and K. Summer. 1975. Hydroperoxide-metabolizing systems in rat liver. Eur. J. Biochem. 1975. 57:503-512.

Sosnovsky, G. and D. Rawlinson. 1971. The chemistry of hydroperoxides in the presence of metal ions. In Organic Peroxides, ed. by D. Swern. New York: Wiley Interscience. p.153-268.

Spector, A. and W.H. Garner. 1982. H_2O_2-induced formation of lens opacity in cultured lens. Fed. Proc. 41:1370-1374.

Srivastava, S. and E. Beutler. 1969a. The transport of oxidized glutathione from human erythrocytes. J. Biol. Chem. 244:9-16.

Srivastava, S. and E. Beutler. 1968b. Cataract produced by tyrosinase and tyrosine systems in rabbit lens in vitro. Biochem. J. 112:421-425.

Tappel, A. 1969. Vitamin E as the biological lipid antioxidant. Vitam. Horm. 20: 493-510.

Turrens, J., Freeman, B., Levitt, J. and J. Crapo. 1982. The effect of hyperoxia on superoxide production by lung submitochondrial particles. Arch. Biochem. Biophys. 217:401-410.

Ursini, F., Maiorino, M. and C. Gregolin. 1985. The selenoenzyme phospholipid hydroperoxide glutathione peroxidase. Biochim. Biophys. Acta 839:62-70.

Weisiger, R. and I. Fridovich. 1973. Superoxide dismutase: organelle specificity. J. Biol. Chem. 248:3582-3592.

Wendel, A. 1980. Glutathione peroxidase. In Enzymatic Basis of Detoxification, Vol. 1, edited by W. Jakoby. New York: Academic Press. p.333-353.

Light Damage

William T. Ham, Jr., Harold A. Mueller, and R. Kennon Guerry

The literature on ocular damage from light is extensive and involves various tissues of the eye as well as many different spectra, exposure times and power levels. This presentation will be confined primarily to retinal damage from prolonged exposures, i.e. greater than one second, to the short wavelengths in the visible spectrum (500–400 nm) and to the near ultraviolet (UV) wavelengths (UVA, 400–320 nm). Exposure at power levels too low to cause thermal damage can result in photochemical damage. There are at least two types of photochemical damage that need to be considered. One type entails primary or initial damage to the photoreceptors brought about by light absorption in the photopigments of the rods and cones. The action spectrum for this type of photochemical damage follows the absorption spectrum of the photopigments, particularly rhodopsin, and was first demonstrated convincingly in albino rats (Noell, et al., 1966). The other type of photochemical insult results in primary and/or initial damage to the retinal pigment epithelium (RPE), brought about by the absorption of short wavelength light (500–400 nm) in the melanin granules located in the apical portion of the RPE (Ham, et al., 1976 and 1978). The action spectrum for this type of photochemical injury approximates roughly the absorption spectrum of melanin, extending into the near ultra-violet (UV) and has been characterized variously as UV damage, photochemical or short wavelength damage, actinic or abiotic damage and/or blue light damage.

The photochemical effects of near UV radiation are similar but more exaggerated than those of blue light damage and include, in addition to injury to the RPE cells, extensive damage to the photoreceptors, especially the cones (Ham, et al., 1982). The absorption of light in the retina also can produce thermal as well as photochemical effects and thermal enhancement of photochemical effects must be considered. The ocular media of the mammalian eye which includes the cornea, the aqueous, the lens and the vitreous humor transmit wavelengths between 400 and 1400 nm (Guerry, et al., 1985). The mature human lens is an excellent yellow filter that protects the retina from much of the blue light and nearly all of the radiation below 400 nm but when the lens is removed, as in cataract surgery, the retina is exposed to near UV as well as additional amounts of blue light.

QUANTUM ASPECTS OF NEAR INFRARED, VISIBLE AND NEAR UV RADIATION:

For the longer wavelengths in the visible and near infrared spectrum vibrational and/or rotational quantum states of excitation predominate over electronic excitational states when radiation is absorbed by molecular systems in aqueous media. Molecules in the ground electronic state which are rotationally and vibrationally excited, usually dissipate their energy back to the surrounding water molecules by radiationless transitions without disruption of their molecular structure. As the surrounding media heats up, the larger macromolecular systems are subject to thermal denaturation. Hydrogen and hydrophobic/hydrophilic bonds are broken and macromolecules lose their tertiary structure, unwind, and become susceptible to cross bondings or linkages that result in polymerization or coagulation with permanent loss of function. This probably corresponds to the mildest form of thermal injury as represented by a minimal retinal lesion requiring 24 hours after exposure to become visible in the fundus camera or ophthalmoscope.

At wavelengths below approximately 500 nm, electronically excited quantum states become more numerous so that photochemical or actinic reactions become more probable. These effects result from extended exposure to short wavelengths in the visible spectrum at power levels too low to produce temperature changes of more than a few degrees Celsius. The transition from thermal to photochemical retinal injury is gradual as wavelength decreases, resulting in an ill-defined mixture of the two types of insult through the spectral interval 600-500 nm. There is a region of thermally enhanced photochemical reactions bridging this gap between predominately thermal and predominately photochemical events.

ACTION SPECTRA OF BLUE LIGHT AND NEAR UV RADIATION:

The action spectrum for minimal retinal damage was determined by Ham, et al., (1976) when they exposed rhesus monkeys to eight laser lines extending from 1064 nm in the near infrared to 441 nm at the blue end of the spectrum. The biological endpoint was the observation of a minimal retinal lesion with the fundus camera at 48 hours postexposure. Retinal irradiance in W/cm^{-2} was determined for each wavelength for exposure times of 1, 16, 100 and 1000 seconds. Retinal sensitivity to minimal injury rose exponentially as the wavelength decreased below 500 nm. Temperature rise above ambient was calculated as a function of exposure time for each of the eight laser lines. Rises less than $10^{o}C$ were assumed to be too low to cause thermal damage, while rises above $20^{o}C$ almost certainly represented irreversible thermal injury. The region between was assumed to represent a mixture of thermal and photochemical toxicity to the retina. For short exposures, 1-10 seconds, wavelengths above 514 nm produced thermal damage, whereas those below 500 nm produced predominately photochemical injury. Wavelengths below 450 nm caused only photochemical damage, even for exposure times as short as 1 s; on the other hand the 1064 nm infrared line of the Nd:YAG laser produced thermal damage even for exposures as long as 1000 s. The corneal power input required to produce a minimal lesion in the monkey retina was three (3) orders of magnitude greater for the 1064 nm infrared line than for the 441 nm blue line.

The action spectrum for minimal retinal damage was extended into the near UV by exposing 3 aphakic eyes of rhesus monkeys to 405, 380, 350 and 325 nm wavelengths produced by a 2500 W xenon lamp optical system with quartz optics and 10 nm wide interference filters (Ham et al, 1982). Transmittance vs wavelength for the cornea, the aqueous and the vitreous with the lens excluded, was calculated for each wavelength using Maher's data for the rhesus monkey (Maher, 1978). When the lens was removed approximately 50% of the near UV at 350 and 325 nm incident on the cornea was transmitted to the retina. Exposure times of 100 and 1000 seconds and retinal image sizes of 500 micrometer diameter were used to correspond to the previous data for the visible spectrum (Ham,et al., 1976). Retinal sensitivity to photic damage continued to increase exponentially in the near UV. Actually at 350 and 325 nm, only 5 J/cm^{-2} was required to inflict a minimal lesion as compared to 30 J/cm^{-2} at 440 nm blue light. This represents a factor of six (6) so that the rhesus retina is 6 times more sensitive to near UV than to blue light.

HISTOLOGICAL CHARACTERIZATION OF PHOTOCHEMICAL/ACTINIC RETINAL LESIONS:

Both the blue light lesion and the near UV lesion have been investigated histologically and ultrastructurally. They appear to be similar except that ultraviolet radiation causes extensive damage to the photoreptor cells, especially the cone ellipsoids which appear to be especially vulnerable, probably due to absorption by the metalloflavoproteins and cytochromes in the mitochondria. Also rhodopsin and the cone visual pigments have strong absorption peaks in the the near UV and this may help to explain why photoreceptor damage is more pronounced for near UV radiation than for short wavelength visible light. In both types of exposure damage to the retinal pigment epithelium (RPE) plays an important and similar role. The blue light lesion has been characterized by histological analysis on 3000 sections from 20 eyes in 10 rhesus monkeys after exposures to 441 nm light at radiant exposures 10% above the threshold for minimal injury (Ham, et al., 1978). Sections were processed at postexposure times ranging from 1 hour to 90 days. No evidence of injury at one hour and one day postexposure was found, except for a few pyknotic nuclei in the outer nuclear layer and an occasional dense cone ellipsoid. At 2 days postexposure, the RPE was inflammed and edematous over 90% of the exposed area (1 mm diameter). The most characteristic feature was depigmentation caused by the agglutination of melanin granules that produced interstices in the curtain of melanin granules normally situated in the apical RPE. A few macrophages containing melanin granules were present in the subretinal space. It seemed quite evident that melanin somehow was involved vitally in the blue light type of actinic or photochemical lesion. Lesions examined at 5-6 days postexposure showed a highly inflammed RPE, often with cellular proliferation and always with marked depigmentation. Numerous macrophages loaded with melanin granules were now visible in the subretinal space and for the first time the outer segments of the photoreceptors above the lesion showed damage and disarrangement. At 10 to 11 days postexposure there was remarkable recovery. The

RPE, though depigmented, showed less proliferation and usually consisted of a single layer of cells; macrophages still persisted in the subretinal space but the outer segments of the photoreceptors seemed less damaged. After 30 days most lesions examined had returned to normal except for depigmentation and macrophages in the subretinal space. At 60 to 90 days there was almost complete recovery. The macrophages had disappeared and depigmentation was only slight. Lesions were barely discernable and had a grainy or speckled appearance similar to that seen in the aging retina.

SOLAR RETINOPATHY:

The characteristic features of the blue light lesion as outlined above resemble closely the clinical events resulting from solar retinitis (Guerry, et al., 1985) as well as the syndrome of foveomacular retinitis in military personnel reported by Marlor, et al., (1973) and the actinic macular retinal pigment degeration in 150 servicemen stationed on a tropical island in the Pacific, as reported by Smith (1944). The recovery phase after blue light exposure is similar to the clinical data on eclipse blindness reported by Penner & McNair, 1966; vision returned to 20/20 after six months in 50% of those affected. After the 1970 total eclipse, 45% of those suffering eclipse blindness returned to normal vision within a few months after exposure, (Hatfield, 1970). The best histopathologic study of the early phase of solar retinopathy in humans is from the remarkable study of Tso, et al., (1975) on 3 patients scheduled for enucleation with extrafoveal malignant melanoma. These patients volunteered to stare at the sun for one hour. The eyes, enucleated 38 to 48 hours postexposure, showed varying degrees of damage to the RPE with edema, irregular pigmentation, and frank necrosis. The photoreceptors appeared normal. In all three patients visual acuity had returned to pre-exposure levels prior to enucleation.

It has been shown in the rhesus monkey that solar retinopathy as produced by a simulated solar source is caused almost entirely by the blue component in the solar spectrum (Ham, et al., 1980). It has also been demonstrated in rhesus monkeys trained to perform a visual task (the Landolt ring technique) that visual performance returns to normal after exposure of the macula to blue light insult, provided the radiant exposure in not too severe (Moon, et al., 1978).

LIGHT HAZARD FROM DIAGNOSTIC AND SURGICAL PROCEDURES IN OPHTHALMOLOGY:

Tso (1977) found disruption of the blood retinal barrier at the RPE in 7 rhesus monkeys subjected to surgical lens extraction but no data were given about light exposure during the surgical procedure nor was there any reference to light exposure as a possible causative agent. Nonetheless, the retina is exposed to high levels of light during diagnostic and surgical procedures. A series of studies have explored the potential hazards to the retina from ophthalmic instrumentation equipped with optical sources, e. g. tungsten, tungsten-halogen, and/or xenon lamps. These include the indirect ophthalmoscope, the slit lamp (biomicroscope) with fundus contact lens, the fundus camera, flourescein angiography, fiberoptic illuminators, overhead surgical lamps and the operating microscope. All of these instruments irradiate the retina with a substantial component of blue light and in some cases with a small component of near UV. For example the average retinal irradiance for the indirect ophthalmoscope is about 69 mW/cm^{-2}; slit lamp biomicroscopy produces retinal irradiances that are 3 times greater than this. Overhead surgical lamps illuminate an area of 5 disk diameters with 24 mW/cm^{-2}. The retinal irradiance of several surgical operating microscopes in current use average about 460 mW/cm^{-2}. (Tso, et al, 1972; Henry, et al., 1977; Fuller, et al., 1978; Robertson, et al., 1979; Calkins, et al., 1979, 1980; Hochheimer, et al., 1979, 1984; Delori, et al., 1980; Mainster, et al., 1983; McDonald, et al., 1983; Guerry, et al., 1985).

Calkins and Hochheimer (1979) have emphasized the potential hazard of light exposure during intraocular surgery. This is particularly true in elderly patients undergoing cataract extraction with intraocular lens implantation. During this procedure, the beneficial yellow filtering effects of the natural lens are removed, and the posterior pole becomes vulnerable to damage from the unfiltered blue light and near UV radiation. Several authors have recently published photographs of fundus lesions observed postoperatively in such patients. The shape of these lesions often corresponds to the image of the illuminating filament in the operating microscope. Without question such lesions are photochemical in nature. Fortunately, in many cases visual loss is not permanent because the blue light lesion is partially reversible if the exposure is not too severe.

LIGHT TOXICITY AND THE AGING RETINA:

The realization that acute or extended exposure to short wave-length light and near UV radiation produces photochemical effects that are toxic to the retina has changed considerably our way of thinking about light damage. Solar retinopathies such as those experienced by military personnel during World War II, as well as eclipse blindness and solar retinitis can now be understood in terms of the blue light lesion. The hazards from extended exposure to blue light may also apply to those who work or play continuously in a bright light environment, e.g. sunbathing, skiing, mountain climbing, yachting as well as welders, surgeons, and nurses in operating rooms, seamen, farmers, lifeguards and airplane pilots. There is increasing speculation that long-term, chronic exposure of the retina to short wavelength light and, in the case of aphakics, to near UV radiation, may accentuate the degenerative and aging processes leading to macula degeneration. The evidence that excessive light exposure increases the metabolic burden of the RPE is convincing (Ham, et al., RPE Working Group, in press). Depigmentation is a significant feature of blue light damage to the macaque retina. This suggests that light accentuates the depigmentation process found in the aging macula. Light exposure also increases the photooxidation of the outer segments of rods and cones, thereby increasing the accumulation of lipofuscin in the RPE and accelerating the collection of debris. This, in turn, further stresses the aging RPE leading to additional exocytosis into Bruch's membrane.

Young (1982) describes the plight of the aging RPE as follows: "In the retina the most prominent signs of senescence are the accumulation of lipofuscin in the cell body of the ganglion cells and the cytoplasm of pigment epithelial cells. Bruch's membrane is also the site of the gradual accumulation of amorphous material, filamemtous and vesicular structures and drusen. Deterioration of the pigment epithelium and Bruch's membrane seems to be the source of most of the degenerative changes in senile maculopathies. Visual cells do not accumulate lipofuscin largely because they shed most of the membranes which they produce in such phenomenal quantities. Compounding the problem of this enormous degradative burden for the pigment epithelium is the large proportion of highly unsaturated fatty acids which are particularly susceptible to peroxidation and abnormal atomic bonding to other molecules. Anything that reduces the efficiency of renewal hastens the progress of senescence. Thus, depriving animals of vitamin E, which protects membranes against oxidation leads to rapid and premature build-up of lipofuscin in the pigment epithelium. Deterioration is also accelerated by the high intensity of radiant energy that is absorbed in the membranes and in the pigment epithelium. The senescent changes are centered on the macula, where the intensity of radiation appears to be the greatest. Accumulation of lipofuscin in the pigment epithelium begins in childhood. Several decades later, the cells are filled to overflowing with the remnants of failing molecular renewal, despite abortive attempts to clear the cytoplasm of debris by extruding it onto Bruch's membrane. According to Hogan, the gradual failure of the function of the retinal pigment epithelium is the principal cause of many forms of degenerative macula disease, especially the senile ones. Senescence of the pigment epithelium may be traced to inefficiency of renewal in a layer of cells subjected to exorbitant demands for molecular degradation."

In the rhesus monkey each retinal rod produces 80 to 90 disks per day; the total outer segment (OS) is replaced every 9 to 13 days. Thus, each pigment epithelial cell must phagocytize and destroy approximately 3000 disks per day (Young, 1971). As Feeney (1973) has pointed out, "The pigment epithelial cell must have a highly developed phagocytic-lysosomal system in order to digest these enormous amounts of exogenous material daily for 70 or more years." Before the advent of modern medicine the average life span was about 35 years. Evolution has not equipped the mammalian retina with the ability to engorge such huge amounts of debris. The consequence is that the aging retina must extrude this undigested material into Bruch's membrane. The process is accelerated by light toxicity, especially in the macular area where light is focussed throughout the day. This accumulation of debris and foreign matter (drusen) in Bruch's membrane is a characteristic feature of the aging retina, leading to age-related macular degeneration, and all too frequently to neovascularization with extrusion of blood vessels into the subretinal space.

THE OXYGEN EFFECT: BASIC MECHANISMS OF PHOTOTOXICITY:

While the basic mechanisms promoting or causing photochemical reactions in the retina are not known specifically there is good reason to believe that the oxygen radicals and reactive molecules O_2^-, H_2O_2, OH· and $^1O_2$ play an important role in producing toxic effects. To test this hypothesis Ham, et al., (1984) exposed macaque monkeys under oxygenation to short wavelength light (435-445 nm) and compared the threshold for retinal damage to that determined under normal conditions. Anesthetized monkeys breathed, through an endotracheal tube with attached gas bag, various ratios of oxygen/nitrogen ranging from 20/80 (air) to 80/20 and 100% oxygen. Arterial blood samples were taken both before and after 30 minutes of breathing a specific mixture and analyzed for PO_2, PCO_2, HCO_3, and pH at a measured rectal temperature. Exposures were performed with a 2500 W xenon lamp optical system equipped with quartz optics. The radiation was peaked at 440 nm by means of a 10 nm interference filter. Exposure duration was electronically controlled at 100 s. The results were definitive. There was a sharp drop in the radiant exposure threshold with increase in the partial pressure of oxygen in arterial blood. Threshold radiant exposure in J/cm^{-2} decreased exponentially with increase in PO_2 in mm of Hg. according to the empirical equation, $H=39.3 \exp(-.0049\ PO_2)$ which was determined from the data on 8 eyes in 4 monkeys breathing various mixtures of oxygen/nitrogen. At a PO_2 of 270 mm of Hg, for example, the threshold was 10.5 J/cm^{-2} as contrasted with 30 J/cm^{-2} for a monkey breathing air.

In unpublished research these same authors have shown that oxygenation reduces the threshold for near UV retinal damage by a factor of three or more. They exposed the aphakic eye (lens surgically removed) of a rhesus monkey to 325 nm UV radiation while the arterial blood oxygen level was elevated to various levels (398, 389, 278 and 139 mm of Hg.). The threshold for retinal damage was reduced from 5.5 J/cm^{-2} for the unoxygenated eye to less than 2 J/cm^{-2} for PO_2's greater than 300 mm of Hg. Histological analysis showed excessive damage to the cone ellipsoids as well as severe injury to the RPE. While these experiments demonstrate that oxygenation enhances retinal sensitivity to blue light and near UV radiation they do not prove necessarily that O_2^-, H_2O_2, OH· and $^1O_2$ are involved.

In another experiment Ham, et al., (1984) fed a rhesus monkey beta-carotene (7.5 mg daily) over a period of 2 years. The threshold for retinal damage from 440 nm light was determined as 30 J/cm^{-2} before beta-carotene feeding began. Breathing a normal ratio of oxygen/nitrogen of 20/80, the threshold was elevated approximately 10%, from 30 to 33 J/cm^{-2}. However, after breathing a ratio of 80/20 oxygen/nitrogen (PO_2 372 mm of Hg) for 30 minutes the threshold was established at approximately 23 J/cm^{-2}, whereas the formula predicted 6.5 J/cm^{-2} for a PO_2 of 372. The protective action of beta-carotene under high arterial oxygen tension implies that singlet oxygen ($^1O_2$) might be the toxic factor, since beta-carotene is known to be a potent scavenger of singlet oxygen. However, beta-carotene can also scavenge many other excited molecular species. More recent experiments have attempted to detect the effects of superoxide dismutase and catalase on the retinal toxicity of blue light (440 nm). These enzymes are specific for the dismutation of O_2^- + O_2^- to H_2O_2 and O_2 and the catalysis of H_2O_2 + H_2O_2 to $2H_2O$ + O_2. They were injected intravenously into monkeys both before and after exposures of the retina to measured radiant exposures. The results were inconclusive, most probably because these macromolecules remain in the circulation for only a short time (6-10 minutes) and would not be expected to penetrate the blood-retinal barrier at the RPE.

Despite the lack of definitive proof that oxygen radicals and excited species are partially if not wholly responsible for the retinal toxicity of blue light and near UV radiation there are a number of reasons for believing that they play an important role. The mammalian retina is unique among body tissues in that during the daytime light is focussed continuously on the macula, a group of cells that are highly oxygenated and metabolically active. The choroidal circulation accounts for the major portion of ocular blood and the choriocapillaris is so structured that a dense matrix of small blood vessels with large surface area is immediately adjacent to Bruch's membrane and the RPE and photoreceptors. The densely packed mitochondria in the ellipsoids attest to the high rate of energy and oxygen consumption needed to synthesize new disks for the outer segments, while the RPE must phagocytize the discarded disks and maintain the flow of nutrients from the choroid to the neural retina. Even without light the photoreceptor and RPE cells would be subject to oxygen toxicity but the combination of light and oxygen enhances the probability of harmful reactions. The retina must maintain a delicate balance between its requirements for oxygen and light and the toxic effects engendered by them.

Photochemical reactions are produced by photons of light exciting a molecular sensitizer, S, to form an initial excited electronic quantum state, the singlet state ^{1}S, that has a short half-life ($< 10^{-8}$ s). The excited molecule can dissipate its energy in three major ways: interaction with the solvent, usually water; emission of a photon (fluorescence); radiationless transition (intersystem crossing-over) to a metastable or triplet state, ^{3}S, that has a long half-life compared to ^{1}S and therefore has more time to react with other molecules. The triplet state, ^{3}S, is generally assumed to be the major pathway leading to most photochemical reactions. According to Foote (1976), the most effective sensitizers are those yielding a long-life triplet state in high quantum yield. The retina is replete with molecular species that could serve as photosensitizers (chromophores), e.g. melanin, retinol, retinal and the photopigments, not to mention the hematoporphyrins, flavins, aromatic hydrocarbons etc. distributed ubiquitously throughout mammalian tissue. In addition to endogenous sensitizers, there are exogenous chromophores, e.g. certain drugs, dyes, food, etc.

There are two major types of photochemical reactions. In type I, redox reactions do not involve oxygen initially and ^{3}S interacts directly with the substrate. In type II, ^{3}S interacts directly with oxygen to produce either singlet oxygen, $^1O_2$, or superoxide anion O_2^-. Photochemical reactions involving oxygen are designated commonly as photodynamical and the majority of such reactions produce singlet oxygen (Foote, 1976). The oxygen molecule in the ground state is a triplet, $^3O_2$, with two unpaired electrons having parallel spins; its reactivity is limited because spin inversion is restricted but the excitation of $^3O_2$ to $^1O_2$ involves spin inversion. $^1O_2$ is more reactive than $^3O_2$ in producing oxidative reactions, especially with polyunsaturated fatty acids (PUFA) which are the major components of cell membranes. The outer segments (OS) of rods and cones are especially vulnerable to $^1O_2$. Singlet oxygen in aqueous media has a half-life of approximatly 3.3 microseconds (Rogers, 1983).

The reduction of oxygen to water requires the removal of four electrons. Univalent single electron transfers can generate superoxide and superoxide radicals dismute spontaneously to form hydrogen peroxide and oxygen. When ferrous ion (Fe^{2+}) is present in aqueous media with O_2^-, H_2O_2 and $^3O_2$, the extremely reactive radical OH· can be generated (Aust, et al., 1982). Fortunately, most respiring cells reduce most of their oxygen by divalent and tetravalent reactions, thereby reducing or in the case of tetravalent reactions removing the production of O_2^-, H_2O_2 and OH·. However, cells do utilize univalent and divalent pathways and there are numerous spontaneous oxidations as well as enzymatic oxidations that can generate free radicals and excited molecules. Mitochondria and phagocytic cells are known to generate O_2 and H_2O_2 which serve as bactericidal agents (Forman & Boveris, 1982) (Babior, 1982). The phagocytosis and digestion of outer segment disks discarded by the rods and cones is a major function of the RPE cell. Part of this process may involve O_2^- and H_2O_2 with subsequent formation of the peroxidized lipids of lipofuscin granules.

A major cause of aging in living systems is thought to be the spontaneous generation of free radicals and reactive molecules by metabolic processes (Harman, 1982; Cutler, 1984). Free radicals also are thought to play an important role in many human diseases including cancer, senile dementia, amyloidosis, cardiovascular disease (atherosclerosis, hypertension) and immune disorders (Harman, 1982).

The retina has developed multiple defense mechanisms against free radicals and photo-induced injury. Molecular turnover or bio- logical renewal is a primary means of removing cellular debris and delaying the onset of senescence; this fundamental process is common to all cells including the RPE and the photoreceptors (Young,1982). Another example common to all mammalian cells is the well known enzyme repair system for DNA. The photoreceptor cells have evolved an unique defense against peroxidation of the polyunsaturated lipid membranes of the outer segments. Rod outer segments renew their disks at a rate of 10% or 90 disks per day; cone outer segments have a slower but appreciable rate. It is the distal tips of the outer segments that have been exposed the longest to peroxidation by light and oxygen; they are shed at a daily rate that is in equilibrium with the synthesis of new disks appearing at the base of the outer segments. Over a lifetime it is estimated that each RPE cell must phagocytize and digest about 100 million disks shed from rods and an unknown number of membranes discarded by cones. As Young (1982) has emphasized, the RPE cells must have a capacity for molecular degradation that far exceed that for any other cells in the body.

Specific substances to combat damage from O_2^-, H_2O_2, OH· and $^1O_2$ are present in the retina. One of these is superoxide dismutase (SOD), an enzyme that catalyzes the dismutation of O_2^- to H_2O_2 and $^3O_2$. Two classes of related enzymes, the catalases and peroxidases, catalyze the reduction of H_2O_2 to H_2O. Catalases, found mainly in liver, kidney and the red blood cell, are efficient for reducing high concentrations of H_2O_2 to H_2O and $^3O_2$ without the requirement of an electron donor. Peroxidases reduce H_2O_2 best at low concentrations and require a co-substrate or hydrogen donor, glutathione (GSH) or ascorbic acid. Glutathione peroxidase, a seleno-enzyme, is found in high concentration in the primate retina.

The dismutases, catalases and peroxidases protect the retina from O_2^- and H_2O_2 and indirectly from OH· by eliminating its precursors but they do not shield it from singlet oxygen which is thought to be a major product of type II photodynamical reactions. Delmelle (1978) has proposed singlet oxygen as a major cause of light damage to the retina. The most destructive role of singlet oxygen is the peroxidation of lipid membranes. Antioxidants supply the main line of defense against lipid peroxidation that can become a chain reaction in membranes. Foremost among naturally occuring antioxidants is vitamin E, alpha-tocopherol, which is relatively concentrated in the retina, especially in the outer segments of the photoreceptor cells where it can intercept radicals, singlet oxygen, and terminate a chain reaction in the lipid bilayer. Vitamin E is also thought to be a scavenger of singlet oxygen and to act synergistically with selenium to protect the cell from free radical damage. Vitamin C, ascorbic acid, is another antioxidant distributed ubiquitously among body tissues. These two vitamins act synergistically, vitamin E acting as the primary antioxidant. The resulting vitamin E free radical, in turn, reacts with vitamin C to regenerate vitamin E. The latter is lipophilic, allowing it to move freely within the lipid bilayer of membranes. Vitamin C is water soluble and available throughout the cytosol.

While vitamin C is a scavenger of singlet oxygen, beta-carotene and the carotenoids in general are probably the best quenchers of singlet oxygen known to occur naturally in primate cells. The number of conjugated double bonds determines primarily the rate of quenching. Carotenes with nine or more double bonds are effective quenchers of singlet oxygen. Quenching is due to electronic energy transfer to the triplet state of beta-carotene with dissipation of this energy to the solvent without loss of molecular structure. Carotenes can also quench the original sensitizer molecule [3]S before it interacts with oxygen or other substrates.

Another potent defense against the toxic effects of oxygen and light is melanin, the major ingredient of the melanin granules situated primarily in the apical portion of the RPE where they are in close apposition to the outer segments of the photoreceptor cells. It is generally assumed that the major role or function of melanin is to shield the retina from scattered light and to convert absorbed photons into harmless heat. While this concept is generally valid melanin may have other functions above and beyond the mere conversion of light to heat. Proctor, et al., (1974) have suggested that melanin is protective at low rates of energy input but that at high rates melanin becomes cytotoxic. Melanin may be equally as protective by absorbing the energy of potentially disruptive excited molecules and free radicals as in absorbing light. Data to support the cytotoxic thesis comes from the histological effects of blue light (Ham, et al., 1978), where 48 hours after exposure the RPE undergoes an inflammatory reaction, accompanied by agglutination of melanin granules with phagocytosis by macrophages. The melanin granules appear to have undergone some type of damage from overexposure to blue light. Possibly melanin granules injured by exposure to blue light become complex granules (melanolysosome and melanolipofuscin granules) similar to those that accumulate in the cytoplasm of the aging RPE cell. Feeney and Berman (1976) suggest that biochemical damage to the RPE by light and/or oxygen be re-examined in view of the free radical character of melanin and its possible role as an electron-transfer agent. Melanin is a heterogenous or random polymer comprising several different monomers coupled by various bond types into an amorphous substance containing stable free radicals and semi-conductor properties. Recent research has shown melanin to have a number of interesting features. When irradiated with light melanin generates free radicals (Cope, et al., 1963); it functions as an efficient electron transfer agent in redox systems (Gan, et al., 1976); it scavenges oxygen in the presence of light, reducing the scavenged oxygen to hydrogen peroxide accompanied by some production of superoxide (Felix, et al., 1978); pheomelanin generates superoxide, hydroxyl radicals and solvated electrons when irradiated in aqueous media by UV radiation and blue light (Chedekel, et al., 1980). Data by Menon and Haberman (1977) indicate that the protective effect of melanin is not entirely due to the absorption of light. They suggest that more attention be paid to the protective and deleterious effects of melanin and that pigment biologists keep an open mind for other possible biological effects that are not presently recognized. Melanin has an affinity for certain drugs and metal ions (Sarna, et al., 1976 and Linquist, 1973). This could be a mixed blessing-- either a storehouse for needed materials or for toxic substances that poison the RPE cell. Whatever its role in light damage melanin should not be considered as just a passive absorber of light and molecular substances.

There is some hope that future experiments with SOD, catalase, glutathione peroxidase, beta-carotene, vitamins E and C, etc. will make it possible to define the role of specific toxic agents like the oxygen free radicals and excited molecules but to prove specificity in vivo will be extremely difficult. Meanwhile, lack of understanding of the basic mechanisms underlying photochemical light damage should not obscure the practical and clinical significance of the oxygen effect.

SUMMARY:

The discussion of light damage is focussed primarily on injury to the primate retina from exposure to short wavelength light (500-400 nm) and near UV radiation (385-325 nm). Quantum aspects of the photochemistry of infrared, visible and near UV radiation are mentioned briefly. The action spectra for retinal damage from blue light and near UV radiation for various exposure times are given, followed by histological characterization of the photochemical blue light lesion in the macaque retina and how it differs from the near UV lesion. Solar retinopathy is explained as a photochemical lesion produced by the short wavelengths in the visible spectrum of the sun. Ophthalmic surgery and diagnostic procedures employ bright sources of short wavelength light that represent a potential hazard to the retina. It is postulated that long-term, chronic exposure to environmental blue light can accelerate the aging process leading to macular degenerations. High levels of arterial blood oxygen increase the sensitivity of the retina to photochemical damage and suggest that oxygen free radicals and excited species are involved in the phototoxicity resulting from exposure to blue light and/or near UV radiation. Beta-carotene protects the retina from blue light damage when high levels of arterial blood oxygen exist during exposure, suggesting that singlet oxygen may be a major toxic factor. The basic mechanisms leading to retinal toxicity are discussed in terms of superoxide, hydrogen peroxide, hydroxyl radical and singlet oxygen. Defense mechanisms against oxygen and light toxicity are given. These include molecular turnover (biological renewal), rod and cone shedding and the enzymes superoxide dismutase (SOD), catalase and glutathione peroxidase. Vitamin E, vitamin C and melanin are also important protective agents against free radicals, excited molecules and other toxic agents.

REFERENCES

Aust, S.D. and Svingen, B.A. 1982. The role of iron in enzymatic lipid peroxidation. In "Free Radicals in Biology", Vol. 5, pp 1-25, edited bu W.A. Pryor, Acad. Press, NY.

Babior, B.M. 1982. The role of active oxygen in microbial killing by phagocytes. In "Pathology of Oxygen", pp 45-48, edited by A.P. Autor, Acad. Press, NY.

Calkins, J.L. and Hochheimer, B.F. 1979. Retinal light exposure from operation microscopes. Arch. Ophthalmol., 97: 2363-2369.

Calkins, J.L., Hochheimer, B.F. and D'Anna, S.A. 1980. Potential hazards from specific ophthalmic devices. Vis. Res., 20: 1039-1053.

Calkins, J.L. and Hochheimer, B.F. 1980. Retinal light exposure from ophthalmoscopes, slit lamps and overhead surgical lamps: An analysis of potential hazards, Invest. Ophthalmol. and Vis. Sci. 19: 1009-1015.

Chedekel, M.R., Agin, P.P. and Sayre, R.M. 1980. Photochemistry of pheomelanin: Action spectrum for superoxide production. Photochem. and Photobiol., 31: 553-555.

Cope, F.W. Sever, R.J. and Polis, B.D. 1963. Reversible free radical generation in the melanin granules of the eye by visible light. Arch. Biochem. Biophys., 100: 171-177.

Cutler, R.G. 1984. Antioxidants, aging and longevity. In "Free Radicals in Biology", Vol. 6 pp 371-428, edited by W.A. Pryor, Acad.Press, NY.

Delmelle, M. 1978. Retinal sensitized photodynamic damage to liposomes. Photochem. and Photobiol., 28: 357-360.

Delori, F.C. Pomerantzeff, O. and Mainster, M.S. 1980. Light levels in ophthalmic diagnostic instruments. Proc. Soc. Photo-Optical Instr. Eng. (SPIE) 229:154-160.

Delori, F.C.Parker, J.S. and Mainster, M.A. 1980. Light levels in fundus photography and fluorescein angiography. Vis. Res. 20: 1099-1105.

Felix, C.C. Hyde, J.S. Sarna, T. and Sealy, R.C. 1978. Melanin photoreactions in aerated media: electron spin resonance evidence for production of superoxide and hydrogen peroxide. Biochem. Biophys. Res. Commun., 84: 335-341.

Foote, C.S. 1976 Photosensitized oxidation and singlet oxygen: consequences in biological systems. In "Free Radicals in Biology", Vol. 2, edited by W.A. Pryor, Academic Press, NY.

Forman, H.J. and Boveris, A. 1982. Superoxide radical and hydrogen peroxide in mitochondria. In "Free Radicals in Biology", Vol.5, pp 65-87, edited by W.A. Pryor, Acad. Press, NY.

Fuller, D.,Machemer, R. and Kingston, R.W. 1978. Retinal damage produced by intraocular fiber optic light. Am. J. Ophthalmol. 85:519-524.

Gan, E.V. Haberman, H.F. and Menon, I.A. 1976. Electron transfer properties of melanin. Arch. Biochem. Biophys., 173: 666-672.

Guerry, R.K. Ham, W.T. and Mueller, H.A. 1985. Light toxicity in the posterior segment. Chap. 37, in Clinical Ophthalmology, edited by Thomas D. Duane, Harper and Row, Phila. PA.

Ham, W. T.,Jr. Mueller, H.A. and Sliney, D.H. 1976. Retinal sensitivity to damage from short wavelength light. Nature 260: 153.

Ham, W.T., Jr., Ruffolo,J.J., Jr.,Mueller, H.A., Clarke, A.M. and Moon, M.E. 1978 Histologic analysis of photochemical lesions produced in rhesus retina by short wavelength light. Invest. Opthalmol. and Vis. Sci. 17: 1029-1035.

Ham, W.T., Jr.,Mueller, H.A. Ruffolo, J.J.,Jr., and Guerry, D,III. 1980. Solar retinopathy as a function of wavelength: its significance for protective eyewear. Chap. in "The Effects of Constant Light on Visual Processes", edited by T.P. Williams and B.N. Baker, Plenum Pub. Corp. , New York, NY.

Ham, W.T., Jr., Mueller, H.A., Ruffolo, J.J., Jr., Guerry, D., III and Guerry, R.K. 1982. Action spectrum for retinal injury from near-ultraviolet radiation in the aphakic monkey. Am. J. Ophthalmol. 93: 299-306.

Ham, W.T., Jr. , (Chrm), Allen, R.G., Feeney-Burns, L., Marmor, M.F., Parver, L.M., Proctor, P.H., Sliney, D.H. and Wolbarsht, M.L. (in press). Retinal Pigment Epithelial Working Group. "The involvement of the RPE in Light Damage." CRC Press. Inc. Boca Raton, FL.

Ham, W.T., Jr.,Mueller, H.A., Ruffolo, J.J., Jr.,Millen, J.E., Cleary, S.F.,Guerry, R.K. and Guerry,D., III. 1984. Basic mechanisms underlying the production of photochemical lesions in the mammaliann retina. Curr. Eye Res. , 3:165-174.

Harman, D. 1982 The free-radical theory of aging. In "Free Radicals in Biology", Vol. 5, pp 255-275, edited by W.A. Pryor, Acad. Press, NY.

Hatfield, E.M. 1970. Eye injuries and the solar eclipse. Sight Saving Rev. 40: 79-86.

Hochheimer, B.F., D'Anna, S.A. and Calkins, J.L. 1979. Retinal damage from light. Am. J. Ophthalmol. 88: 1039-1044.

Hochheimer, B.F. 1981. A possible cause of chronic cystic maculopathy, the operating microscope. Ann. Ophthalmol. 13: 153-155.

Lindquist, N.G. 1973. Accumulation of drugs on melanin. Acta. Radiologica. Suppl., 325: 5-84.

Maher, E.F. 1978. Transmission and absorption coefficients for the ocular lens implantation. Ophthalmology 90: 945-950.

158

McDonald, H.R. and Irvine, A.R. 1983. Light-induced maculopathy from the operating microscope in extracapsular cataract extraction and intraocular lens implantation. Ophthalmology 90: 945-950.

Marlor, R.L., Blair, B.R.Preston, F.R. and Boyden, D.G. 1973. Foveomacular retinitis, an important problem in military medicine: epidemiology. Invest. Ophthalmol., 12: 5-16.

Menon, I.A. and Haberman, H.F. 1977. Mechanisms of action of melanins. Brit. J. Dermatol., 97: 109-112.

Moon, M.E.,Clarke, A.M., Ruffolo, J.J., Jr., Mueller, H.A. and Ham, W.T., Jr. 1978. Visualperformance in the rhesus monkey after exposure to blue light. Vis. Res. 18: 1573-1577.

Noell, W.K.Walker,V.S., Kang, B.S. and Berman, S. 1966. Retinal damage by light in rats. Invest. Ophthalmol. 5: 450-473.

Penner, R. and McNair, J.N. 1966. Eclipse blindness. Am J. Ophthalmol. 61: 1452-1457. Proctor, P., McGinness, J. and Corry, P. 1974. A hypothesis on the preferential destruction of melanized tissues. J. Theor. Biol., 48: 19-22.

Robertson, D.M.and Ericksin,G.J. 1979. The effect of prolonged indirect ophthalmoscope on the human eye. Am. J. Ophthalmol. 87: 519-524.

Rogers, M.A.J. 1983. Time resolved studies of 1.27 micrometer luminesence from singlet oxygen generated in homogeneous and microheterogeneous fluids. Photchem. and Photobiol., 37: 99-103.

Sarna, T., Hyde, J.S. and Swartz, H.M. 1976. Ion-exchange in melanin: an electron spin resonance study with lanthanide probes. Science 192: 1132-1134.

Smith, H.E. 1944. Actinic macula retinal pigment degeneration. U.S. Naval Med. Bulletin 42: 675-680.

Tso, M.O.M., Fine, B.S. and Zimmerman, L.E. 1972. Photic maculopathy produced by the indirect ophthalmoscope. Am. J. Ophthalmol. 13: 686-692.

Tso, M.O.M. and LaPiana, F.G. 1975. The human fovea after sungazing. Trans. Am. Acad. Ophthalmol. and Otolaryngol. 79: 788-795.

Tso, M.O.M. and Shih, C.Y. 1977. Experimental macular edema after lens extraction. Invest. Ophthalmol. and Vis. Sci. 16: 381-392.

Young, R.W. 1971. The renewal of rod and cone outer segments in the rhesus monkey. J. Cell Biol., 49: 303-315.

Young, R.W. 1981. A theory of central retinal disease. Chap. in "Future Directions in Ophthalmic Research", edited by M.L. Sears, Yale Univ. Press, New Haven, CT.

Young, R.W. 1982. The Bowman Lecture: Biological Renewal: Applications to the Eye. Trans. Ophthalmol. Soc. UK, 102: 42-73.

Studies on Biochemical Mechanisms of Retinal Degeneration

Robert E. Anderson, Rex D. Wiegand, Laurence M. Rapp,
Maureen B. Maude, Muna I. Naash, and John S. Penn

Our laboratory is investigating biochemical mechanisms of certain forms of retinal degeneration which specifically affect the photoreceptor cells. One experimental model we use is the albino or pigmented rat retina exposed to constant illumination. Our strategy has been to expose animals to moderate light levels for periods up to five days and analyze for changes in the components of the photoreceptor rod outer segments (ROS) that may result from or contribute to the observed degeneration. Results obtained to date suggest that lipid peroxidation occurs within ROS membranes during light-induced retinal degeneration. In this chapter, we will review briefly results of some of our studies and propose a means by which photoreceptors are protected against lipid peroxidation. Other reviews on this subject have been published by our group (Anderson, et al., 1983; Wiegand, et al., 1984).

In our first set of experiments (Wiegand, et al., 1982; Wiegand, et al., 1983), some albino rats were exposed to 100-120 foot-candles of light for up to three days, while others were maintained on low levels (10-15 foot-candles) of cyclic light (12hL:12hD). Animals were sacrificed, retinas removed, and ROS membranes prepared by discontinuous sucrose flotation. A portion of the membranes was analyzed by polyacrylamide gel electrophoresis to determine if constant light exposure resulted in any gross changes in proteins of these membranes. Another portion of the membranes was extracted with organic solvents for lipid analysis. No gross changes in protein composition or molecular weight distribution were noted in animals maintained in constant light, so we focused our attention on possible changes in the membrane lipids.

In animals maintained for three days in constant illumination, there were no quantitative changes in diacylglycerol and free fatty acid content of ROS lipids. However, thin-layer chromatography of ROS phospholipids showed a loss of phosphatidylserine and phosphatidylethanolamine relative to phosphatidylcholine (Wiegand, et al., 1982). Since the former two phospholipids contain most of the polyunsaturated fatty acids in outer segments, and phosphatidylcholine comparatively less, we thought that there may be a specific loss of polyunsaturated fatty acids in the membranes of animals exposed to constant illumination. This was indeed found to be the case. Molecular species analysis revealed a specific loss of those species containing two polyunsaturated fatty acids (Wiegand, et al., 1984). When expressed as nanomoles of fatty acid per milligram of ROS membrane protein, there was a selective loss of docosahexaenoic acid (22:6ω3) compared to the other fatty acids (Wiegand, et al., 1982; Wiegand, et al., 1983). Surprisingly, the other major polyunsaturated acid, arachidonic acid (20:4ω6), was not significantly reduced.

Spectrophotometric measurement of ROS lipid extracts showed an increase in absorbance at 233 nanometers in animals exposed to constant illumination as compared to controls maintained at low levels of cyclic light (Wiegand, et al., 1983). This absorbance is typical of conjugated dienes which are formed during lipid peroxiation. Thus, the results of these early experiments on albino rats suggested that lipid peroxidation may be involved in photoreceptor degeneration due to constant illumination. Subsequent light damage studies were conducted on pigmented animals whose pupils were dilated with atropine. Similar changes in polyunsaturated fatty acids were found in ROS from pigmented animals following constant illumination (Wiegand, et al., 1985).

The concurrance of lipid peroxide formation and retinal degeneration has been observed in other experimental paradigms in addition to constant light exposure. Earlier studies by Hiramitsu et al. (Hiramitsu, et al., 1974) had shown that intravitreal injection of lipid hydroperoxides in rabbits resulted in a retinal degeneration. Armstrong et al. (1982) subsequently confirmed this finding by showing irreversible eletroretinographic (ERG) amplitude decrease following lipid hydroperoxide injections in rabbits. We have generated lipid hydroperoxides in vitro by intravitreal injection of ferrous sulfate. Ferrous ions are known to produce free radicals which may then initiate lipid peroxidation. In our first series of experiments (Rapp, et al., 1982) frogs were injected with 10 microliters of an aqueous solution of 20 mM ferrous sulfate or an equal volume of 20 mM sodium sulfate as a control. Animals were placed immediately in the dark and electroretinography carried out over a twenty-four hour period. In the animals injected with ferrous sulfate, a rapid loss of both a- and b- waves of the ERG was observed. There was a 50% reduction of the amplitudes of both wave forms by one hour and complete extinction of the ERG after four hours. No ERG change as found in the eyes injected with sodium sulfate. Light and electron microscopy revealed the rapid deterioration of photoreceptor cells, with the earliest changes noted in the outer segments. Surprisingly, there were few changes in the other retinal layers. The retinal pigment epithelium quickly became engorged with retinal debris, and large oil droplets could be seen in greater abundance than usual in the pigment epithelium.

As observed in light-damaged rat ROS, there was an increase in the lipid hydroperoxides in ROS from the ferrous ion-injected frog eyes, compared to control eyes, at four and twenty-four hours after injection. Also, there was a selective loss of 22:6ω3 relative to ROS membrane protein. Palmitic acid (16:0), a saturated fatty acid which does not undergo lipid peroxidation, was unchanged in ROS isolated four or twenty-four hours following ferrous ion injection. Thus, as was observed for light damage, there was evidence for lipid peroxidation in membranes of photoreceptor cells with ferrous ion-induced degeneration.

More recent experiments by Wiegand et al. (1985) in pigmented rats have shown that intravitreal injection of ferrous sulfate leads to an increase in lipid hydroperoxides, selective loss of photoreceptor cells, and predictable loss of function measured by electroretinography.

Evidence to date from our laboratory (Wiegand, et al., 1984; Anderson, et al., 1985; Wiegand, et al., 1982; Wiegand, et al., 1983; Wiegand, et al., 1985; Rapp et al.,1982, Wiegand et al., 1985) and others (Hiramitsu, et al., 1974; Armstrong, et al., 1982; Kagan, et al., 1973; Kagan, et al., 1981; Shevedova, et al., 1979; Yagi, et al., 1977; Hiramitsu, et al., 1976) shows that lipid peroxidation is found in certain types of retinal degeneration. To establish a causal rather than an associative relationship, however, is not an easy task. Simply measuring lipid changes in degenerating membranes will not provide a definitive answer. We have attempted to test the relationship between lipid peroxidation and retinal degeneration using another approach. If photoreceptor cells are indeed susceptible to peroxidative damage, then it is likely that some mechanism(s) exists to protect the cells from peroxidation. Protection of cells is usually mediated through low molecular weight anti-oxidants and enzymatic detoxification of oxidation products. If lipid peroxidation is involved in light damage, then alteration of the level or distribution of these compounds may change the susceptibility of the retina to degeneration.

A prevelant antioxidant in biological tissues is vitamin E, which terminates the chain reaction of lipid peroxidation. Vitamin E deficiency has been shown to result in retinal degeneration in primates (Hayes, et al., 1974), dogs (Riis, et al., 1981), and rats (Robison, et al., 1979), and the possibility exists that the degeneration resulted from an increased vulnerability to lipid peroxidation. Other small molecules which provide anti-oxidant protection in cells are ascorbic acid (vitamin C) and glutathione. The concentrations of vitamin E, ascorbic acid, and glutathione are quite high in the retina, and it is reasonable to presume that they provide antioxidant protection for the retina. This possibility will be discussed in more detail later.

The retina also contains high levels of superoxide dismutase (Hall, et al., 1975), an enzyme that converts superoxide radicals to hydrogen peroxide. Hydrogen peroxide, which is also toxic to cells, can be converted to water and molecular oxygen by catalase and glutathione peroxidase. Both of these enzymes are present in the retina (Stone, et al., 1982; Armstrong, et al., 1981). Although we have not specifically analyzed for catalase, we have determined that photoreceptor outer segments contain relatively high levels of glutathione peroxidase (Naash, et al., 1984). A strategy employed by our laboratory and several others has been to attempt to alter the levels of putative antioxidant molecules and/or some of the enzymes that detoxify reactive oxygen species in the retina. With a few exceptions that will be discussed later, these studies have yielded only inconclusive evidence regarding a mechanism for light damage to the retina. Although long-term vitamin E deficiency results in a retinal degeneration, several groups (Stone, et al., 1979; Paglia, et al., 1967) including our own (Rapp, et al., 1985), have failed to demonstrate that vitamin E deficiency leaves the

animal more susceptible to damaging effects of light.   While
there are obvious differences between retinal changes due to long-
term deprivation and those due to constant light, the fact
nevertheless remains that animals "deficient" in vitamin E do not
appear to be more susceptible to the damaging effects of constant
illumination.   However, in studies of this type, there is always a
question of the extent of vitamin E deficiency.   For example, we
lowered the vitamin E content of the photoreceptors by 75%,
decreasing the amount of vitamin E per moleclule of 22:6ω3 from
1:500 to 1:2000.   Since vitamin E functions as a terminator of
lipid-free radicals, it is still quite possible that one in two-
thousand is adequate to provide protection against damaging effects
of light.   Attempts by our laboratory to reduce the level of
glutathione in photoreceptors by inhibiting the activity of the
enzyme glutathione reductase, have also produced a similar result.
There is no apparent enhancement of light damage in animals where
glutathione reductase activity has been significantly altered.
Changes in the retinal levels of vitamin C have been shown to be
associated with altered susceptibility to light damage.   Tso et al.
(1984) showed that constant illumination decreased the levels of
reduced ascorbic acid in primate retinas.   Li et al. (1985) and
Organisciak et al. (1985) showed that rats injected intraperi-
toneally with ascorbic acid were protected from damage by constant
illumination.   There was less retinal damage in animals receiving
ascorbic acid, as determined by morphometric analysis of the outer
nuclear layer thickness (Li, et al., 1985) Furthermore, those
animals receiving ascorbic acid injections also contained higher
levels of 22:6ω3 (Organisciak, et al., 1985), indication that
ascorbic acid protected against the loss of this polyunsaturated
fatty acid which otherwise occurs during constant illumination.

   Parts of the following scheme were proposed in 1968 by Tappel
(1968) as a means by which low molecular weight antioxidants might
protect against lipid peroxidation.

Toxic aldehydes

$$R\cdot \underset{RH}{\overset{LH}{\rightleftarrows}} \underset{L\cdot}{\overset{LOOH}{\rightleftarrows}} \underset{LOO\cdot}{\overset{}{\rightleftarrows}} \underset{Vit\ E\ (red)}{\overset{Vit\ E\ (ox)}{\rightleftarrows}} \underset{Vit\ C\ (ox)}{\overset{Vit\ C\ (red)}{\rightleftarrows}} \underset{GSH}{\overset{GSSG}{\rightleftarrows}} \underset{NADP}{\overset{NADPH}{\rightleftarrows}}$$

O_2

Lipid peroxidation is initiated in a bilayer by the attack of lipid molecules (LH) by the free radical R·. Unless some radical terminator interferes, the peroxy radical (LOO·) can react with LH in a chain reaction that produces hundreds of molecules of lipid hydroperoxides (LOOH). Vitamin E reacts with peroxy radicals to form a lipid hydroperoxide and oxidized E. This vitamin E radical is more stable than the lipid alkyl (L·) or peroxy radical, and its formation is therefore the favored reaction. Furthermore, vitamin E (ox) will not continue the chain reaction.

Packer et al. (1979) have shown that the unpaired electron on oxidized vitamin E can be transferred to ascorbic acid, resulting in regeneration of the reduced form of vitamin E in the lipid bilayer and oxidized ascorbic acid. This reaction has also been observed in the in vitro oxidation system of Pryor and colleagues (Pryor, et al., 1985). Further, oxidized ascorbic acid reacts nonenzymatically with glutathione to yield ascorbic acid and oxidized glutathione (Naash and Anderson, unpublished observation). Oxidized glutathione can be regenerated to the reduced form via the enzyme glutathione reductase.

Applying Tappel's scheme to the situation in the retina, we propose the following: vitamin E present in the outer segment membrane scavenges lipid radicals and is oxidized to a less reactive radical. Ascorbic acid acts at the lipid-aqueous interface to reduce vitamin E and is itself oxidized. Glutathione reduces ascorbic acid and is regenerated by the enzyme glutathione reductase. All three antioxidant molecules are present in fairly high concentrations in outer segments, and glutathione reductase activity has also been demonstrated in outer segments (Naash and Anderson, unpublished observations).

As mentioned earlier, the mechanism of photoreceptor degeneration from constant illumination is not known. While there is evidence suggesting that lipid peroxidation is involved, it is far from conclusive. Additional experiments of the the type described above appear necessary in order to draw definitive conclusions regarding a role of lipid peroxidation in light-induced retinal degeneration. However, several lines of evidence do suggest that the retina is a target for lipid peroxidation. The high oxygen flux through the non-vascular portion of the retina to the mitochondria of the photoreceptor cells and the great preponderance of polyunsaturated fatty acids (more than 50% of the total fatty acids) in these membranes make them especially vulnerable to peroxidation. To deal with this high oxidation potential, the retina has built an elaborate protective mechanism. Outer segments contain the enzymes superoxide dismutase, glutathione peroxidase, glutathione reductase, and catalase. All of these enzymes are known to contribute to detoxification of active oxygen radicals and/or species. Furthermore, these cells also contain high levels of vitamin E, vitamin C, and glutathione, which have also shown anti-oxidant capabilities. It will be intriguing over the next several years to investigate the role of this elaborate anti-oxidant defense system in providing protection against drug- and environmentally-induced retinal degenerations.

ACKNOWLEDGEMENTS

This work was supported by grants from the National Eye Institute/National Institutes of Health, Research to Prevent Blindness, Inc., National Retinitis Pigmentosa Foundation, National Society to Prevent Blindness, the Retina Research Foundation, and Alcon Research Institute.

REFERENCES

Anderson, R.E., L.M. Rapp, and R.D. Wiegand (1983). Lipid peroxidation and retinal degeneration. Current Eye Res. 3:223-227.

Armstrong, D., T. Hiramitsu, J. Gutteridge, and S. E. Nilsson (1982). Studies on experimentally induced retinal degeneration. 1. Effect of lipid peroxides on electroretinographic activity in the albino rabbit. Exp. Eye Res. 35:157-171.

Armstrong, D., G. Santangelo, and A. Cornell (1981). The distribution of peroxide regulating enzymes in the canine eye. Curr. Eye Res. 1:225-242.

Hall, M. and D. Hall (1975). Superoxide dismutase of bovine and frog rod outer segments. Biochem. Biophys. Res. Commun. 67:1199-1204.

Hayes, K.C. (1974). Retina degeneration in monkeys induced by deficiencies of vitamin E or A. Invest. Ophthalmol. 13:499-510.

Hiramitsu, T.Y., Y. Majima, Y. Hasegawa, and K. Hirata. (1974). Role of lipid peroxide in the induction of retinopathy by X-irradiation. Acta. Soc. Ophthalmol. Jap. 78:819-825.

Hiramitsu, T., Y. Hasegawa, K. Hirata, I. Nishigaki, and K. Yagi (1976). Lipoperoxide formation in the retina in ocular siderosis. Experientia 32:1324-1325.

Kagan, V.E., I.Y. Kuliev, V.B. Spirichaev, A.A. Shvedova, and Yu.P. Kozlov (1981). Accumulation of lipid peroxidation products and depression of retinal electrical activity in vitamin E-deficient rats exposed to high-intensity light. Bull. Exp. Biol. Med. 91:144-147.

Kagan, V.E., A.A. Shvedova, K.N. Novikov, and Y.P. Kozlov (1973). Light-induced free radical oxidation of membrane lipids in photoreceptors of frog retina. Biochim. Biophys. Acta 330:76-79.

Li, Z.-Y., M.O.M. Tso, H.-m. Wang, and D.T. Organisciak (1985). Amelioration of photic injury in rat retina by ascorbic acid: a histopathologic study. Invest. Ophthalmol. Vis. Sci. 26:1589-1598.

Naash, Muna I. and R.E. Anderson (1984). Characterization of glutathione peroxidase in frog retina. Curr. Eye Res. 3:1299-1304.

Organisciak, D.T., H.-m. Wang, Z.-Y. Li, and M.O.M. Tso (1985). The protective effect of ascorbate in retinal light damage of rats. Invest. Ophthalmol. Vis. Sci. 25:1580-1588.

Packer, J.E., T.F. Slater, and R.L. Willson (1979). Direct observation of a free radical interaction between vitamin E and vitamin C. Nature 278:737.

Paglia, D.E. and W.N. Valentine (1967). Studies on the quantitative and qualitative characterization of erythrocyte glutathione peroxidase. J. Lab. Clin. Med. 70:158-169.

Pryor, W.A., M.J. Kaufman, and D.F. Church (1985). Autoxidation of micelle-solubilized linoleic acid. Relative inhibitory efficiencies of ascorbate and ascorbyl palmitate. J. Organic Chem. 50:281-283.

Rapp, L.M., L.A. Thum, A.P. Tarver, and R.D. Wiegand (1985). Vitamin E and taurine deficiency: effects on the retina in cyclic light-maintained and light-damaged rats. Invest Ophthalmol. Vis. Sci., Supple, 26:131.

Rapp, L.M., R.D. Wiegand and R.E. Anderson (1982). Ferrous ion-mediated retinal degeneration: Role of rod outer segment lipid peroxidation. In Problems of Normal and Genetically Abnormal Retinas, edited by R. Clayton, J. Haywood, H. Reading an A. Wright, Academic Press, pp. 109-119.

Riis, R.C., B.E. Sheffy, E. Loew, et al. (1981). Vitamin E deficiency retinopathy in dogs. Am. J. Vet. Res. 42:74-86.

Robison, W.G., Jr., T. Kuwabara and J.G. Bieri (1979). Vitamin E deficiency and the retina: Photoreceptor and pigment epithelial changes. Invest. Ophthalmol. Vis. Sci. 18:683-690.

Shvedova, A.A., A.S. Sidorov, K.N. Novikov, et al. (1979). Lipid peroxidation and electric activity of the retina. Vision Res. 19:49-55.

Stone, W.L. and E.A. Dratz (1982). Selenium and non-selenium glutathione peroxidase activities in selected ocular and non-ocular rat tissues. Exp. Eye. Res. 35:405-412.

Stone, W.L., M.L. Katz, M. Lurie, et al. (1979). Effects of dietary vitamin E and selenium on light damage to the rat retina. Photochem. Photobiol. 29:725-730.

Tappel, A.L. (1968). Will antioxidant nutrients slow aging processes? Geriatrics 23:97.

Tso, M.O.M., B.J. Woodford, and K.W. Lam (1984). Distribution of ascorbate in normal primate retina and after photic injury: a biochemical, morphological correlated study. Curr. Eye Res. 3:181.

Wiegand, R.D., J.G. Jose, L.M. Rapp, and R.E. Anderson (1984). Free Radicals and Damage to Ocular Tissues. In Free Radicals in Biology and Aging, edited by D. Armstrong, et al., Raven Press (New York), in press.

Wiegand, R.D., C.D. Joel, L.M. Rapp, J.C. Nielsen, M.B. Maude, R.E. Anderson (1985). Polyunsaturated fatty acid and vitamin E in rat rod outer segments during light damage. Invest. Ophthalmol. Vis. Sci., in press.

Wiegand, R.D., N.M. Giusto and R.E. Anderson (1982). Lipid changes in albino rat rod outer segments following constant illumination. In Problems of Normal and Genetically Abnormal Retinas, edited by R. Clayton, J. Haywood, H. Reading and A. Wright, Academic Press, pp. 121-128.

Wiegand, R.D., L.M. Rapp, and R.E. Anderson. Ferrous Ion-Induced Retinal Degeneration: Biochemical Changes in Photoreceptor Membranes. Presented to the Association for Research in Vision and Ophthalmology, Sarasota, Florida, May 6-10, 1985.

Wiegand, R.D., N.M. Giusto, L.M. Rapp and R.E. Anderson (1983). Evidence for rod outer segment lipid peroxidation following constant illumination of the rat retina. Invest. Ophthalmol. Vis. Sci. 24:1433-1435.

Yagi, K., S. Matsuoka, H. Ohkawa, N. Ohishi, Y. Takeuchi, and H. Kakai (1977). Lipoperoxide level of the retina of chick embryo exposed to high concentration of oxygen. Clin. Chim. Acta 80:355-360.

Environmental Factors in Cataractogenesis in RCS Rats

Helen H. Hess, J. S. Zigler, Jr., T. L. O'Keefe, and J. J. Knapka

Royal College of Surgeons (RCS) rats have an inherited disease of the retinal pigmented epithelium (RPE), which shows a greatly reduced capacity to phagocytize the shed terminal portion of the outer segment of the rod photoreceptor cell (Mullen & LaVail, 1976). This failure in the symbiotic relationship of these two cells results in accumulation of outer segment membrane debris at the interface between the cells, and the lack of normal nutriture and membrane renewal leads to the death of all the rod photoreceptors by about 65 days in pink-eyed dystrophics and by 100 days in black-eyed dystrophics (LaVail & Battelle, 1975).

Since its discovery almost half a century ago, the RCS rat has been suggested to be a model for some type of retinitis pigmentosa, but as yet no variety has been found with the early onset and debris accumulation characteristic of the pathophysiological picture in the RCS rat. Another aspect of the RCS disease is the occurrence of cataracts, which first brought these rats to the attention of the scientific community (Bourne et al., 1938a). Studies at that time indicated the cataracts were unpredictable in incidence and time of maturation, in contrast with the retinal degeneration, which was inherited in an autosomal recessive fashion (Bourne et al., 1938b). Recently we found that when studied by slit lamp biomicroscopy, 100% of pink-eyed and black-eyed dystrophics developed bilateral posterior subcapsular opacities (PSO) by 7 to 8 postnatal weeks (Hess et al., 1982; 1983). This meant that the onset of the cataracts was as predictable as that of the retinal degeneration. Posterior subcapsular cataracts are associated with many types of human retinitis pigmentosa and with gyrate atrophy of the choroid and retina (Heckenlively, 1982; Kaiser-Kupfer et al., 1983).

The link between retinal degeneration and PSO may be secondary and an hypothesis has been developed and evidence presented that toxic aldehyde products from peroxidized polyunsaturated lipids of rod outer segments (ROS) may initiate PSO (Zigler et al., 1983a; Goosey et al., 1984; Zigler & Hess, 1985). Several factors predispose degenerating ROS to peroxidation: (1) the very high levels of polyunsaturated lipid present in their membranes (Anderson & Andrews, 1982); (2) the rich supply of oxygen present in the retina; and (3) the disruption of the tissue itself, which in various systems has been shown to be a critical factor leading to lipid peroxidation. In addition, rhodopsin bleaching in the disrupted ROS debris layer will free retinaldehyde, which has been suggested to be a sensitizer that can act to increase photo-oxidative processes, perhaps by generating singlet oxygen ($^1O_2$). In vitro, retinaldehyde has been shown to be capable of giving rise to this highly reactive species of oxygen, which can oxidize poly-unsaturated fatty acids of phospholipids, as well as vitamins E and C which ordinarily protect them (Kagan, 1973; Delmelle, 1977: Krasnovsky & Kagan, 1979). In pink-eyed dystrophics, the rate of conversion of retinaldehyde to retinol is decreased; at 37 days of age, after a 10 min. exposure to 150 footcandles of bleaching light, only 45% of freed retinal is changed to retinol, as compared to 80% in control rats (Delmelle et al., 1975). This abnormally high concentration of free retinaldehyde in bleached retinas may be an important factor in the extreme susceptibility of the retina of RCS dystrophics to light damage, a vulnerability well-documented by LaVail & Battelle (1975). A third factor tending to increase oxidative damage may be the entrance of macrophages into the retina, where they are seen first at 22-26 days in the outer nuclear layer and outer plexiform layer, and later in the debris zone at 28-30 days (LaVail, 1979). Activated macrophages generate active forms of oxygen (superoxide radical, hydrogen peroxide, hydroxyl radical, and perhaps singlet oxygen), which participate not only in intracellular digestive functions of macrophages but also in their recently recognized secretory functions (Nathan et al., 1980).

Counting macrophages in the retina is not feasible, but recently we found that in rats with retinal degeneration the vitreous cortex can be isolated readily (O'Keefe et al., 1985). We counted the macrophages/sq. mm. area of vitreous cortex at 18-130 postnatal days in (1) pink- and black-eyed dystrophics and congenic control, reared in dim cyclic light (1-4 footcandles inside the cage); in (2) pink-eyed dystrophics reared in darkness; and in (3) pink-eyed dystrophics that had been reared in cyclic light, and fed different diets that have been shown to influence the incidence of mature cataracts (observed over a 1 yr period). The studies to be described are consistent with the idea that the number of macrophages at different ages may correlate with increased retinal degeneration and with increased levels of associated peroxidized lipids, whose toxic aldehyde end products have been found damaging to the lens (Zigler et al., 1983a,b; Zigler & Hess, 1985).

Dark rearing and dark eye pigmentation in dystrophic rats were associated with fewer macrophages in the vitreous cortex, and mature cataracts were prevented (O'Keefe et al., 1985). Dark rearing prevented onset of the PSO, an indication that light initiates the cataractous change. Diets containing factors that inhibit or avoid peroxidizing conditions were associated with fewer mature cataracts.

METHODS

Animals used were (1) pink-eyed, tan-hooded dystrophic and congenic control RCS rats (rdy/rdy, p/p and rdy$^+$, p/p, respectively); and (2) black-eyed, black-hooded dystrophic and congenic control RCS rats (rdy/rdy, p+ and rdy$^+$/p$^+$, respectively). Cyclic light reared rats were maintained in plastic cages under a 12 hr on, 12 hr off light schedule. The cages were placed on metal racks that cut off much of the light (light inside the cages was 1-4 footcandles).

Dark reared rats were progeny of pregnant females transferred from cyclic light into a dark room approximately one week prior to expected delivery. The dark room had a secondary door that insured that no light entered the room. Work in the dark room was carried out under red light.

In the dark rearing study, rats were fed the NIH-07 natural ingredient diet for laboratory rodents (Knapka et al., 1974). In the cyclic light studies, several diets were used, as noted in the text. These comprised: (1) the NIH-07 diet; (2) the NIH-42 diet, containing only casein as an animal product; (3) a commercial natural ingredient diet, #3500, from Charles River (prepared by Agway); (4) a purified diet, AIN-76, of the American Institute of Nutrition (AIN Report, 1977); and (5) various supplements to these diets. The general compositions of these diets and of sunflower kernels used as a supplement are shown (table I), as well as the concentrations of natural ingredients (table II) and the vitamins and minerals of special interest in the diets (table III). More detailed compositions have been published (Hess et al., 1981, 1985).

Methods of observation included slit lamp biomicroscopy, dissection microscopy, low power microscope photography, and light microscope histopathology (hematoxylin and eosin staining).

To obtain the vitreous cortex for counting cells, rats of different ages (18-130 postnatal days) were killed with carbon dioxide and the eyes excised and placed in Dulbecco's phosphate buffered saline, pH 7.4. The eye was placed on a saline-moistened gauze square on the stage of a dissecting microscope (10-70X). The globe was cleaned of external tissue, especially at the area of the optic nerve, and the eye was divided into anterior and posterior halves by cutting about 1 mm posterior to the ora serrata with scissors having 8 mm long blades. The two halves were kept together and transferred to a plastic petri dish full of saline. In dystrophic rats with advanced retinal degeneration, the vitreous cortex was loose and often separated cleanly from the retina when the two halves of the eye were gently pulled apart. In young dystrophics and in normal controls the vitreous cortex is strongly attached and careful dissection is necessary to recover it. First it was detached from lens and ciliary body. Then, the rim of the cup was grasped with one pair of forceps while the other pair was closed inside the cup, just short of the retina. The vitreous cortex could usually be pulled out of the cup with this maneuver. The isolated web-like structure was floated in a petri dish containing saline in which a few crytals of Brilliant Green had been dissolved. After staining, the cells were counted. The web was laid on the grid of a hemocytometer and a coverslip (cut from a thin coverslip for mounting sections), 6 x 6mm., was used to trap a single layer of cells. Cells were counted in a 1 sq. mm. area using the conventions for counting white blood cells.

The normal vitreous cortex is attached by collagen fibrils to the basement membranes of the retinal glial cells (Muller cells), ciliary epithelium, and lens capsule. Postmortem autolysis releases the vitreous from the basement membranes of the Muller cells, and so does the pathology of retinal degeneration, both in RCS rats and in light damage. The vitreous cortex can then readily be removed by

TABLE I
COMPOSITIONS OF DIETS AND SUPPLEMENTS (% of diet weight)

Nutrient	AIN-76	NIH-07	NIH-42	#3500	Sunflower Kernels
Fat	5.0	5.3	4.1	5.3	49.3
Protein	20.0	24.0	18.7	24.9	28.3
Ash	3.5	6.7	7.1	7.7	3.5
Fiber	5.0	3.7	5.0	3.3	3.0
Carbohydrate	65.0.	57.5	62.3.	49.1	20.0

severing its other attachments (Balasz & Denlinger, 1984). The mean number of cells per mm.sq. (or per cu./mm.) in congenic control RCS rats was similar to the numbers in bovine vitreous at the retinal equator.

Lenses within the anterior half of the dissected eye were observed while floating in saline, using a Bausch and Lomb Stereozoom or a Zeiss Stereomicroscope SR. The lens was later removed from the anterior segment and observed from anterior and posterior viewpoints.

EFFECT OF DIET ON MATURE CATARACT INCIDENCE IN PINK-EYED RCS RATS

A number of investigators have found the RCS rat strains difficult to breed satisfactorily. Initially, we fed the NIH-07 diet, which had been developed as a total diet for conventionally reared laboratory mice and rats and contained recommended concentrations of all known nutrients for rats (Knapka et al., 1974). When they were fed this diet both dystrophic and control strains produced offspring, but many of the pups failed to survive to weaning and often were cannibalized by their dam. A commercial diet (Charles River R-M-H #3500) was then fed because it approximated a diet reported to be adequate for these strains; the major difference was a lower fat concentration. Since the #3500 diet seemed to result in better survival of pups, we added more fat by supplementing with sunflower kernels. Two months after feeding the #3500 diet supplemented with sunflower kernels, entirely satisfactory reproduction and survival of young to weaning were obtained (Hess et al., 1981). As the progeny of rats fed this diet grew to adulthood (being fed the diet), we noted that they failed to develop grossly visible cataracts, which had been seen at 3-12 months in rats fed NIH-07 and many other diets. This suggested a nutritional-environmental-genetic interaction in the RCS disease.

TABLE II
NATURAL INGREDIENTS IN DIETS (% of diet weight)

Ingredient Added	AIN-76	NIH-07	NIH-42	#3500
Casein (or dried skim milk)	20.0	5.0	3.0	+
D,L-methionine	0.3	--	--	--
Fishmeal	--	10.0	--	+
Soybean Meal	--	12.2	--	+
Isolated soy protein	--	--	2.0	--
Alfalfa meal	--	4.0	--	+
Ground corn	--	24.5	28.0	+
Corn gluten meal	--	3.0	6.6	--
Ground wheat	--	23.0	21.5	+
Wheat middlings	--	10.0	--	+
Wheat bran	--	--	6.0	--
Whole ground oats	--	--	23.5	--
Brewers dried yeast	--	2.0	2.0	+
Dry molasses	--	1.5	--	--
Soybean oil	--	2.5	--	--
Corn oil	5.0	--	1.75	--
Animal fat & BHA	--	--	--	+
Meat & bone meal	--	--	--	+
Sucrose	50.0	--	--	--
Cornstarch (purified)	15.0	--	--	--
Salt	--	0.5	0.6	+
Dicalcium phosphate	--	1.25	2.35	--
Ground limestone	--	0.5	--	--
Calcium carbonate, pure	(Min.Mix)	(Min.Mix)	1.20	(Min.Mix)

All diets had vitamin and mineral mixes added.

The RCS pink-eyed dystrophics fed NIH-07 diet predictably developed slit lamp-detectable posterior subcapsular opacities at 7 to 8 weeks of age. This occurred bilaterally in 100% of all pink-eyed rats, and in black-eyed dystrophics as well (Hess et al., 1982; 1983). As had been reported by LaVail et al., (1975), one fourth of the pink-eyed tan-hooded dystrophics developed mature cataracts by 3-12 months, but in black-eyed dystrophics this occurred in less than 3%. These results suggested that light was a factor in maturation of the cataract.

Slit lamp examination of the pink-eyed dystrophics fed the sunflower kernel supplemented #3500 diet showed that the initial PSO had been minimized. This commercial diet, however, had a "closed" formula; that is, the amounts of the natural ingredients are not revealed to the purchaser (table II and IV). Consequently, the question of which nutrient or nutrients were effective in preventing the cataracts could not be pursued with this diet. The types of diets available for nutritional investigations are summarized in Figure 1, as well as specific diets we have used. An "open" formula natural ingredient diet such as NIH-07 gives more opportunity to determine the origin and amounts of nutrients or toxic factors being

TABLE III

CONCENTRATIIONS OF VITAMINS AND MINERALS OF INTEREST IN THE DIETS

	AIN-76	NIH-07	#3500	Sunflower Kernels
(IU/Kg)				
Vitamin E	50.	37.	75.	350.
Vitamin A	4000.	15,000.	30,800.	507.
(% of Diet)				
Calcium	0.52	1.25	1.5	0.099
(mg/Kg)				
Iron	35.	255.	289.	66.0
Zinc	30.	50.	97.7	49.9
Copper	6.	16.	9.0	16.7
Selenuim	0.1	0.23	0.17	0.77

fed, but the complexity of all natural ingredients is a serious problem in nutritional research. We chose, therefore, to use the purified diet recommended by the American Institute of Nutrition (AIN-76), which contains only highly refined or synthetic ingredients (table IV). All nutrients in this diet are present at levels optimal for growth and not greatly in excess of recommendations by the Committee on Animal Nutrition, National Research Council. The AIN diet has been compared with the NIH-07 diet as the sole source of nutrition for growth of normal rats (NRC, 1978).

The diet most suited to rigorous study of nutritional effects is the chemically defined diet (table IV), which is made with purified nutrients, such as amino acids, triglycerides, essential fatty acids, salts and vitamins. The expense of this diet is much greater than that of the purified diet and it would be used only if reseach objectives required a diet free of complex ingredients.

Controlled studies were conducted to compare the effects of feeding the purified diet AIN-76 alone or with various additions or substitutions, and the NIH-07 and #3500 diets with and without additions (table V).

The fewest cataracts were observed with (1) AIN diet alone or with added sunflower kernels; with additional calcium or with a doubling or tripling of the mineral mix; (2) NIH-42, a diet that contains no animal product except casein (3%), and (3) #3500 diet with added sunflower kernels. Interestingly, the high incidence of mature cataracts that occurred with #3500 (without sunflower kernels) was reduced by autoclaving (a change from 29.0 to 7.5%). All these lower incidences of cataracts with the diets mentioned are significantly different from the incidences seen with diets

TABLE IV
TYPES OF DIETS

1. Chemically defined diets: Chemically pure compounds, such as amino acids, sugars, triglycerides, essential fatty acids, inorganic salt, and vitamins are used to prepare these diets. (Not used in the present study.)

2. Purified diets: Diets formulated with refined ingredients are designated as purified diets. Casein is usually a source of protein; sugar or starch is a source of carbohydrate; vegetable oil or lard is a source of fat; and a form of cellulose is a source of crude fiber; chemically pure inorganic salts and vitamins are added. (AIN-76 diet)

3. Natural ingredient diets: These diets are formulated with appropriately processed whole grains (wheat, corn, oats) or commodities that have been subjected to limited amounts of refinement (fishmeal, soybean meal, wheat bran).

4. Open formula diets: Diets manufactured in accordance with a readily available quantitative ingredient formulation. NIH-07 is an open formula natural ingredient diet.

5. Closed formula diets: Diets manufactured and marketed by commercial institutions which consider the quantitative ingredient composition of the diet privileged information. Diet #3500 of Charles River (Agway) is a closed formula natural ingredient diet.

6. Supplemented diets: A total diet that includes a manufactured diet and an additional ingredient, i.e., #3500 diet pelleted with ground sunflower kernels.

All diets have vitamins and minerals at levels equal to or greater than recommended by the National Research Council.

#3500 and NIH-07 alone, by the Chi Square test, Student's t test, or the t test for paired samples (data from monthly observations of cataract incidence).

The AIN diet with two-fold the usual mineral mix not only resulted in an absence of mature cataracts in a group of 59 rats, but also yielded a higher percentage of lenses that were clear by slit lamp examination (68% as compared with 29% for the AIN diet with the standard mineral mix, and 50% for the AIN diet with three times the mineral mix). The numbers of rats studied in these diets are as indicated in Table IV, except for the AIN diet, in which the number was 69 (the second of two sets of rats studied with this diet). The p value was less than 0.001 for the difference between the AIN diet (standard mineral mix) and the AIN diet with twice the mineral mix; it was less than 0.01 for the difference between AIN and AIN with three-fold the mineral mix, as well as between the two diets with increased mineral mix. Statistics were calculated using the method for comparing two percentages in two large samples, base on the Null Hypothesis and a binomial distribution (Bailey, 1959).

One of the major ingredients present in NIH-07 and #3500 that we chose to examine was fishmeal. Two batches of fishmeal were used

TABLE V

EFFECT OF VARIOUS DIETS ON INCIDENCE OF MATURE CATARACTS IN PINK-
EYED RETINAL DYSTROPHIC RCS RATS BY ONE YEAR OF AGE (REARED AT 1-4
FOOTCANDLES OF LIGHT)

Diet	# of Rats	% of Rats with Mature Cataracts in 1 year
AIN (Am. Inst. Nutr,. 1976)	120	1.7
AIN + 25% sunflower kernels	82	0.0
AIN, 1.5% Calcium(3 x AIN Ca)	68	1.5
AIN, 3.0% Calcium(6 x AIN Ca)	62	0.0
AIN, 2 x Mineral Mix	59	0.0
AIN, 3 x Mineral Mix	41	4.9
AIN with Fishmeal to give 20% protein, replacing casein		
Batch A	60	20.0
Batch B	63	38.1
NIH-07	93	26.9
NIH-07 + 25% sunflower kernels	83	18.1
NIH-07 + Vitamin E (150 IU/Kg) (total E = 187 IU/Kg)	65	30.8
NIH-42 (natural ingredient diet, lacking fishmeal)	55	3.6
#3500, not autoclaved	55	29.0
#3500, autoclaved	67	7.5
#3500 + 25% sunflower kernels	81	4.9

in two different experiments. The fishmeal replaced the casein in
the AIN diet. Batch A fishmeal resulted in an incidence of 20% of
rats with mature cataracts, while batch B increased the incidence to
38%, the highest rate in any experiment. Different batches of
fishmeal are well known to vary in composition. These results
suggest strongly that some factor in fishmeal could contribute to
the cataractogenesis seen with NIH-07 and #3500, and lacking in NIH-
42 and the AIN purified diet which prevent the cataracts. We have
not yet determined whether autoclaved fishmeal loses its
cataractogenic effect. The NIH-07 diet and #3500 diets were fed
without autoclaving. Because fishmeal contains a high concentration
of calcium the experiments of increasing the calcium concentration
to three and six fold the level in the AIN diet were conducted.
Calcium is also present at higher than recommended levels in NIH-07
and #3500. Clearly, additional calcium was not cataractogenic, even
though the six-fold increase had an adverse effect on growth of the
rats. A concomitant control experiment was conducted by increasing
the mineral mix by factors of two and three, without significant
effect on the incidence of mature cataracts.

One of the constituents found in high concentration in sunflower
kernels is vitamin E, and rats fed the #3500 + sunflower kernel diet
were found to have serum tocopherol concentrations three-fold normal

at 50 days of age, the time that the PSO are first detectable by slit lamp (Hess et al., 1981). This showed that the vitamin E was being effectively absorbed. Since NIH-07 was our standard diet that predictably permitted the cataracts to occur, we chose to supplement it with an amount of vitamin E calculated to be present in #3500 when supplemented with sunflower kernels. We found that this diet preserved the morphological intactness of ROS for a longer time (42 days) than NIH-07 alone, and the time of onset of mature cataracts was delayed. Eventually, however, the number of mature cataracts was the same as with NIH-07 (30.8%). This is not surprising, since the phagocytic defect of the RPE had not been eliminated. This experiment demonstrated that vitamin E alone could not account for the effect of sunflower kernels in reducing the incidence of mature cataracts to 18% when used as a supplement with NIH-07 (Hess et al., 1985), although this was a reduction significant (P < 0.01) by the t test for paired samples.

CORRELATION OF CHANGES IN PINK-EYED RCS DYSTROPHIC RETINA AND RPE WITH THOSE IN THE LENS:

Table VI gives a timetable for correlating pathological and chemical changes. Figure 1 outlines the general proposed mechanism of cataract formation. The first changes in the ROS are seen after they have reached their full length and the process of shedding of the tip and phagocytosis by the pigment epithelium should begin (12 days). As shown by Goldman & O'Brien (1978) the RCS rat RPE has only 5% of the normal capacity for phagocytizing the ROS, and debris builds up rapidly. By 18 days, the ROS layer has doubled in thickness and by 22 days the rod inner segments (RIS) and nuclei show signs of damage. At this time the RPE tight junctions have been noted to become permeable to extracellular tracers (Caldwell & McLaughlin, 1983). Also, macrophages first appear in the inner retinal layers at this age. By 27 days the debris has reached maximal thickness, macrophages have appeared in the debris, and at this time the thiobarbituric acid test (TBA) performed on the vitreous indicated the presence of peroxidized material (Zigler & Hess, 1985). Studies of rubidium efflux from cultured lenses of RCS rats showed an increase in passive permeability beginning at 40 days (Zigler & Hess, 1985). The PSO detected by slit lamp appeared at 50 days, and at the same time vitreous cortex macrophages showed a

TABLE VI
CORRELATION OF RETINA-RPE AND LENS CHANGES IN PINK-EYED RCS RAT

Postnatal Days	Retina-RPE	Lens
12	First changes in ROS	
15	Eyes open	
18	ROS layer doubled in thickness	
22	1st changes in rod RIS & nuchei RPE tight junctions permeable Macrophages appear in inner layers	
27	Maximal debris Macrophages in debris	TBA reactivity in vitreous
40	Debris becoming dis-organized	Rb effux first increased
50		PSO first noted by slitlamp Macrophages elevated in
56	Outer retina capillaries are becoming permeable	vitreous cortex. 100% of rats have PSO bilaterally
60	Most rod nuclei gone, debris less	Maximal Rb leakiness
70	Many RPE cells are abnormal in shape, junctions are lost and pinocytotic vesicles appear	
80		Most PSO are internalized Rb efflux is normalized Macrophages decreased in vitreous cortex
110	Debris gone in posterior retina	In some lenses, PSO are not internalized; Rb efflux not normalized; these lenses later become mature cataracts

sudden 4-5 fold rise in number, declining again by about 80 days (O'Keefe et al., 1985). By 56 days, the capillaries of the outer retina are becoming permeable (Essner et al., 1979; Gerstein & Danltzker, 1969), and at this time all pink-eyed, as well as black-eyed dystrophics have bilateral slit lamp-detectable PSO (Hess et al., 1982, 1983). By 60 days most of the rod nuclei have disappeared and the debris has declined; at this time also the lens in culture shows maximal leakiness to Rb, indicating lens membrane damage, which impairs the ability to regulate cation levels. A similar damage was demonstrated in normal lenses which were exposed to degenerating ROS in culture (Zigler et al., 1983a). By 70 days, many RPE cells were abnormal in shape, had lost their junctions entirely, and pinocytotic vesicles were seen in the cells (Calwell &

Figure 1

MECHANISMS OF CATARACT FORMATION IN PINK-EYED RETINAL DYSTROPHIC RCS RATS

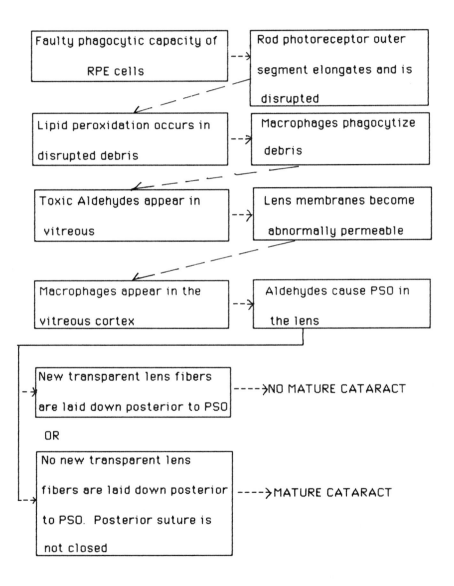

McLaughlin, 1983). By 80 days, in approximately 75% of the rats fed standard rodent diet, the PSO has become internalized, with clear lens fibers being laid down posterior to it, as though a toxic agent producing it had ceased to be present (Hess et al., 1983, 1985); at this time the rubidium efflux had returned to normal in most lenses. We have hypothesized that the toxic agent is an aldehyde product of peroxidation of the polyunsaturated fatty acids of photoreceptor phospholipids (Zigler & Hess, 1985). Toxic aldehydes combine rapidly with amino groups of many tissue components, such as proteins, lipids and nucleic acids. The effects of such aldehydes in various systems may include an increase in cell permeability, inhibition of many enzymes and interference with cell multiplication and growth. A probable site for the action of the toxic aldehydes on the lens is the highly vulnerable bow region where lens cells are multiplying and elongating, and which is especially near to the periphery of the retina. In the rat, however, which has a very large lens relative to the volume of the eye, all areas of the retina are near to the lens, such that toxic agents from the degenerating retina would have only a short distance to travel to reach the lens.

In one fourth of the pink-eyed dystrophic rats the PSO does not become internalized, but remains subcapsular, while the posterior suture fails to close. These opacities have a characteristic appearance by slit lamp and can be predicted to become mature cataracts. These appear to be the same lenses in which the Rb efflux is not normalized. The first mature cataracts may appear as early as 77 days. The time of onset of the mature cataracts as well as their incidence can vary greatly with diet (table V; Hess et al., 1985).

FACTORS IN CATARACTOGENESIS AND PREVENTION IN PINK-EYED RCS RATS

General Factors Promoting Lipid Peroxidation: (1) High concentration of polyunsaturated fatty acids (PUFA) in tissue lipids. (2) Tissue disruption. (3) Presence of photosensitizing molecules. (4) High oxygen tension. (5) Ozone. (6) Presence of metal catalysts (iron, copper). (7) Low defenses against destructive oxygen species and free radicals: singlet oxygen ($^1O_2$) superoxide anion radical (O_2^{-}); and hydrogen peroxide (H_2O_2); and (8) Low defenses against lipid peroxy radicals and lipid hydroperoxides, resulting from effects of destructive oxygen species and free radicals on PUFA.

Special Factors Promoting Lipid Peroxidation in Pink-eyed RCS Rat ROS:

1. ROS membranes have a higher content of long-chain polyunsaturated fatty acids than other membranes that have been examined (Anderson & Andrews, 1982). They have up to one third of their phospholipid fatty acids as docosahexaenoic acid (C22.6) whose 6 double bonds are especially vulnerable to peroxidation.

2. At 18 days, the rhodopsin-containing ROS debris is 2 times the thickness of the normal layer, and the thickness continues to increase until 27 days (Dowling & Sidman, 1962).

3. As the debris accumulates, the membranes become disrupted, which renders them more susceptible to lipid peroxidation. By 25 days, only about half the rhodopsin is in surviving visual cells; by 35 days, only 25%; and by 80 days, none (Delmelle et al., 1975).

4. The conversion of retinaldehyde (released from bleached rhodopsin) to retinol by the cation of alcohol dehydrogenase (NADP/NADPH dependent) is decreased, owing presumably to enzyme damage. When rats 37 days old were exposed to bleaching green light (490-480nm) of 150 footcandles intensity, only 45% of freed retinal was converted to retinol, compared to 80% in normal animals (Delmelle et al., 1975).

5. The transfer of retinol from ROS debris to RPE is slowed increasingly from 17 days onward. Retinol itself is damaging to cell membranes, and is also susceptible to oxidation (Delmelle et al., 1975).

6. The regeneration of rhodopsin is impaired after 17 days of age. By 37 days, regeneration after 1 hr exposure to strongly bleaching light is only 30% of normal after 6 hr in darkness (Delmelle et al., 1975).

7. The dystrophic RCS rat is extremely vulnerable to light damage to the retina. Exposure at 27 days of age for less than 0.5 hr to strongly bleaching light, destroys almost all remaining visual cells. Even exposure to cyclic light of 2 footcandles for 8 days beginning at 20 days of age increases the rate of the ERG -wave decline (Delmelle et al., 1975).

8. This extreme vulnerability to light damage may be related to the impaired handling of retinaldehyde which may act as a photosensitizer capable of generating singlet oxygen, as has been

demonstrated to occur in vitro (Delmelle, 1977,1979; Krasnovsky & Kagan, 1979). Singlet oxygen can attack the polyunsaturated fatty acids of ROS membrane phospholipids, as well as vitamins E and C that ordinarily protect them. It can also oxidize retinaldehyde, as well as retinol.

9. Hydroperoxides have been demonstrated in vitro in retinas and ROS exposed to light, and the action spectrum of the process is that of rhodopsin (Kagan et al, 1973). They have also been detected in ROS exposed in vivo to constant illumination (100-125 footcandles) for 3 days (Wiegand et al., 1983; Anderson et al., 1984).

10. Loss of the blood retina barrier in the dystrophic rat may in some way influence lipid peroxidation and other pathological changes. This process begins at 3 weeks of age and steadily worsens over a 7 month period, when not only the RPE barriers but also the barriers of the capillaries of the outer retina have been lost (Essner et al., 1979; Caldwell & McLaughlin, 1983).

General Defenses Against Effects of Destructive Oxygen Species and Free Radicals on PUFA:

1. The first line of defense is from chemical compounds and enzymes that destroy singlet oxygen, superoxide anion radical and hydrogen peroxide (Fridovich, 1976; Bellus, 1978; Witting, 1980).

A. Singlet oxygen can be scavenged by vitamins E and C and by glutathione (GSH). An hypothesis is that beta-carotene could scavenge singlet oxygen more efficiently and protect the other antioxidants (Burton & Ingold, 1984), but this has not been tested in tissues such as retina.

B. Superoxide anion radical can be destroyed by superoxide dismutase, a copper-zinc metalloenzyme, before this radical can give rise to other oxidizing species which attack PUFA.

C. Hydrogen peroxide is destroyed by catalase (an iron-containing metalloenzyme), without generating free radicals.

2. The second line of defense is from antioxidant compounds and enzymes that react with lipid peroxy free radicals and lipid hydroperoxides (Witting, 1980).

A. Lipid peroxy free radicals (LOO*) can be trapped by vitamins E and C and by GSH, with a chain-breaking effect on PUFA oxidation. An hypothesis is (Burton & Ingold, 1984) that at low oxygen tensions, as in blood capillaries, beta-carotene also can act to tie up lipid peroxy free radicals, forming a resonance- stabilized carbon-centered radical (LOO-beta-Car*), which by a cross-termination reaction, will yield non-radical products. This has not been tested in tissues such as retina.

B. Lipid hydroperoxides (LOOH) are scavenged by being converted to hydroxy fatty acid (LOH) by (1) the selenium-containing GSH peroxidase, and (2) a non-selenium-containing peroxidase, (probably GSH Transferase).

The nutrients of importance in these defenses are vitamins E and C, GSH, and minerals zinc, copper, selenium, and iron. Vitamin E is normally present at a high concentration in ROS (Dilley & McConnell, 1970) and is necessary for their structural integrity (Farnsworth & Dratz, 1976; Robison et al., 1979). Albino rats maintained from weaning on diets deficient in vitamin E and selenium have been shown to have a 75% reduction in RPE phagocytosis in some areas of the retina, together with a 20-34% decrease in photoreceptor cells and accumulation of ROS debris between RPE and photoreceptors (Katz, et al., 1982); these defects were especially prominent when the rats were made deficient in other antioxidants (chromium and sulfur amino acids) in addition to vitamin E and selenium. Vitamin E is also necessary to prevent degeneration of cones and rods in the monkey retina, especially in the macula (Hayes, 1974). The rat can synthesize vitamin C and it is not added to the diets. The extent to which dietary carotenoids could provide antioxidant defense in the retina remains to be determined.

PREVENTION OF CATARACTS BY DARK REARING; EVIDENCE FROM ENUMERATION OF VITREOUS CORTEX MACROPHAGES AND OBSERVATIONS OF ISOLATED LENSES AT 20-85 DAYS

The lenses of dark reared pink-eyed dystrophics were removed at intervals from 48-85 days, placed in saline and observed under a dissecting microscope (O'Keefe et al., 1985). No PSO were seen during that time period. It seems likely that if kept in the dark these rats would never develop mature cataracts. Histophathological studies showed no sign of opacity in such lenses, and the retinas retained photoreceptor cell nuclei for a longer period of time than did those of cyclic light reared dystrophics of the same age, as reported previously (Dowling & Sidman, 1962; LaVail & Battelle, 1975). The total absence of light greatly reduced the rate of

TABLE VII
CORTICAL VITREOUS CELLS AT 50-53 POSTNATAL DAYS IN RCS DYSTROPHICS FED
DIFFERENT DIETS AND REARED IN DARKNESS OR CYCLIC LIGHT OF 1-4 FOOTCANDLES

Strain	Rearing	Cells/sq.mm.		(n)	% of Rats with Mature Cataracts	Diet*
		Mean	+ SEM			
Pink-eyed	Darkness	106.	+ 6.11	(20)	0.0	NIH-07
Black-eyed	Cyclic light	155.4	+ 4.17	(24)	0.0	#3500 + SFK
Pink-eyed	Cyclic light	359.1	+ 16.75	(52)	26.9	NIH-07
Pink-eyed	Cyclic light	317.6	+ 21.68	(16)	4.9	#3500 + SFK
Pink-eyed	Cyclic light	464.1	+ 36.02	(13)	38.1	AIN + Batch B Fishmeal

*See Methods, and tables I-III.
 See table V (mature cataracts occurred between 3-12 months of age).
Differences in cortical vitreous cells significant at the P< 0.001 level:
 (1) Values for pink-eyed rats reared in darkness or of black-eyed rats
reared in cyclic light compared with each other and all other groups.
 (2) Value for pink-eyed rats fed AIN + fishmeal, compared with all groups
except cyclic light reared pink-eyed rats fed NIH-07; the latter comparison
was significant at the P< 0.01 level.
Values for cyclic light reared pink-eyed rats fed diets NIH-07 and 3500 +
SFK were not significantly different (P< 0.1).
n = number of eyes studied.

retinal degeneration and the concentration of macrophages in the
vitreous cortex (table VII; figure 2) and prevented formation of PSO
in pink-eyed dystrophics. The macrophage counts were compared with
results obtained previously in cyclic light reared rats.

Figure 2 depicts the mean values for cells/sq. mm. of vitreous
cortex in (1) dark reared pink-eyed, tan-hooded dystrophics from 20-
85 days; (2) cyclic light reared black-eyed, black-hooded
dystrophics from 20-130 days; (3) cyclic light reared pink-eyed,
tan-hooded dystrophic RCS rats from 28-127 days; and (4) pink-
eyed,tan-hooded and black-eyed, black-hooded congenic controls from
17-83 days (O'Keefe et al., 1985). The cyclic light for rearing the
rats was 1-4 footcandles inside the cages. The values for the
black-eyed dystrophics have been interconnected by a dotted line for
clarity, since some values were close to or superimposed upon those
for pink-eyed dystrophics. Pink-eyed dystrophics, both dark-reared
and cyclic light reared, were fed NIH-07 diet and black-eyed
dystrophics were fed #3500 + sunflower kernel diet. All congenic
control rats were fed NIH-07.

Figure 2:   Each point for the cyclic light reared pink-eyed
dystrophics represents 3-20 eyes; the cyclic light reared black-eyed
dystrophics, 3-8 eyes; and the dark reared pink-eyed dystrophics; 2-
13 eyes.

Figure 2

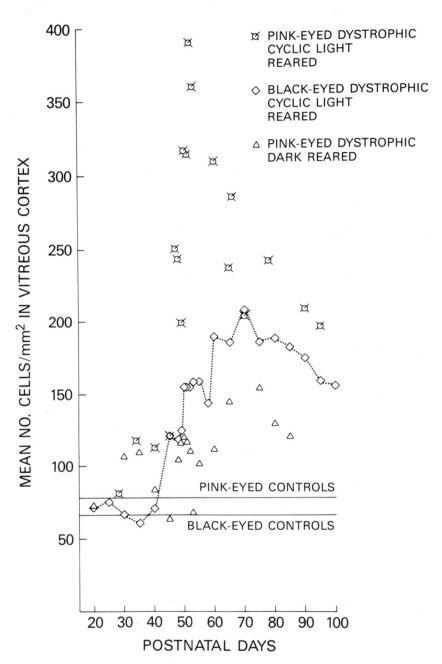

TABLE VIII

CORTICAL VITREOUS CELLS IN PINK- AND BLACK-EYED CONGENIC CONTROL RCS
RATS REARED IN CYCLIC LIGHT

Strain	Postnatal Days	Cells/sq. mm., Mean+SEM (n)
Pink-eyed Control	19-83	78.2 + 3.08 (44)
Black-eyed Control	17-73	66.2 + 3.08 (19)

Results for pink-eyed controls were significantly different from
those in black-eyed controls (P < 0.01)
(n) is number of eyes studied

Values for pink- and black-eyed congenic controls (figure 2 and
table VIII) were nearly constant and averaged 78 cells/mm^2 for the
pink- and 66 cells/mm^2 for the black-eyed rat. The black-eyed
control value was significantly less than that of the pink-eyed
control (P<0.01), suggesting that even at low light levels retinal
light damage was occurring in the pink-eyed rat, although this was
too slight to be evident by light microscopic examination of the
retina. The susceptibility of pink-eyed normal rats to light damage
is well established (Noell, 1966; Kuwabara & Gorn, 1968), and damage
occurs even at low light intensity.

Table VI summarizes the results in the dystrophic strains for
the most outstanding period, 50-53 postnatal days, when the PSO
become visible by slit lamp or by dissecting microscopy. The means,
standard errors of the means (SEM) and P values were calculated
using Student's t test. The table also shows the diets fed the
rats, and the incidence of mature cataracts after 12 months in a
group of rats fed the diet.

The number of macrophages in vitreous cortex was significantly
lower in dark reared pink-eyed dystrophics, in comparison not only
with cyclic light reared pink-eyed rats, but also with black-eyed
cyclic light reared dystrophics. The rate of retinal degeneration
has been noted to be slower in dark reared pink-eyed rats, just as
it is in black-eyed cyclic light reared rats. Since the macrophage
concentration reached its most abrupt and highest peak in the pink-
eyed dystrophic reared in cyclic light and its lowest levels in the
pink-eyed dystrophics reared in darkness, the macrophage
concentration may be a measure of the amount of retinal degeneration
that has been occurring at some previous time. A large, single peak
in the pink-eyed rat reared in cyclic light may be related to the
similar rate of degeneration in all sectors of the retina, thus
producing higher concentrations of products that may stimulate

entrance of macrophages into the vitreous cortex. The lower concentrations and double peaks of cells in the dark reared pink-eyed and the cyclic light reared black-eyed dystrophics may be related to the different rates of degeneration in various sectors of the retina. The rate of degeneration in the inferior periphery is the same in black-eyed and pink-eyed dystrophics reared in cyclic light, while there is a delay of 10 days in the central retina and 31 days in the superior periphery in the black-eyed rat (LaVail & Battelle, 1975). The fact that the macrophage count in the dark reared pink-eyed dystrophic was lower that that in the black-eyed dystrophic reared in cyclic light, not only at 50-53 days but also at 60-80 days was a surprise. It has been thought that black pigmentation fully protects the retina from light damage, except when the pupil is dilated by drugs and the animal is subjected to much more intense lighting than 1-4 footcandles. However, the present results suggest that cyclic light even of low intensity may produce a small degree of light damage to the retina in black-eyed dystrophics and that dark rearing might reduce the concentration of cells in their cortical vitreous. The black-eyed dystrophics showed an onset of the PSO at 7-8 weeks, just as the pink-eyed dystrophics did when reared in cyclic light. The concentration of macrophages at 50-53 days in the black-eyed dystrophic, was significantly greater than in the dark-reared pink-eyed dystrophic, and that may indicate a difference in retinal degeneration (presumably in the inferior periphery) sufficient to initiate the PSO. Dark rearing black-eyed dystrophics probably can be expected to prevent the appearance of PSO just as it does in the pink-eyed dystrophic.

AIN + Batch B fishmeal resulted in the highest cell count and the highest percentage of mature cataracts in pink-eyed, tan hooded animals. We have not found that any of the diets affected the incidence of mature cataracts in the black-eyed dystrophics, including the AIN + Batch B fishmeal. This suggests that pigmented iris and RPE protect the dystrophic rat from mature cataract development and point to light as a powerful factor in maturation.

SUGGESTED MECHANISMS OF PREVENTION OF LENS OPACITIES BY DIET AND BY DARK REARING

Diets that influenced the onset or reduced the incidence and maturation of the cataracts in pink-eyed dystrophic RCS rats had the following characteristics:

    1.   Contained no unpurified complex natural ingredients (AIN-76).

    2.   Contained limited amounts of natural ingredients of animal origin (AIN-76; NIH-42).

    3.   Contained sunflower kernels as a supplement. This natural ingredient contains high concentrations of vitamin E and of selenium, as well as moderately high concentrations of zinc and copper, all factors conferring antioxidant properties on the diet. Oil seeds contain a low content of iron, which can have a pro-oxidant effect in high concentrations, and supplements of sunflower kernels would dilute the level of iron in diets NIH-07 and #3500.

    4.   Had been autoclaved. Charles River (Agway) diet #3500 (table V).

    Elimination of fishmeal from a diet would be expected to reduce the risk of mature cataract. NIH-42 was a natural ingredient diet lacking fishmeal, and the incidence of mature cataracts was low. The reduction of cataract incidence after autoclaving of #3500 diet (table V) points to a volatile ingredient as the factor or one of the factors involved. Possible volatile ingredients would include vitamin A (very high in #3500 diet, table III), nitrosamines, and toxic metals such as mercury and cadmium. Nitrosamines and toxic metals are not present in NIH-07 diet in amounts known to be harmful to normal rats, but with breakdown of the blood-retina barrier in dystrophic RCS rats even low concentrations could be significant. A low level of methylmercury in fish has been suggested to be a risk factor in human cataractogenesis (Lane, 1984). Low levels of activity of superoxide dismutase and glutathione peroxidase in erythrocytes were correlated with active cataractogenesis. Increase in dietary selenium and zinc would have some protective effect against mercury and cadmium.

    Further work will be needed to determine (1) whether excess vitamin A, or very low levels of nitrosamines or toxic metals are involved in increasing the percentage of mature cataracts in RCS rats, and (2) whether the antioxidant metals selenium, zinc, and copper are the nutrients which account for the greater clarity of lenses in rats fed AIN with twice the concentration of the mineral mix.

    The effects of dark rearing in preventing onset of the PSO suggest that if rhodopsin bleaching is avoided and retinaldehyde is not set free, there may be less toxic aldehyde lipid degeneration products formed to harm the lens. This would be consistent with the hypothesis that the lipid peroxidation may arise from retinaldehyde generation of singlet oxygen, which is a highly reactive excited state of oxygen. Little has been done to test for generation of singlet oxygen in vivo, and use of singlet oxygen scavengers more effective than vitamin E, such as beta-carotene, could be of interest. The numbers of macrophages in the retina and in the cortex of the vitreous is also low in dark reared dystrophic RCS rats, and toxic oxygen species such as superoxide anion and hydrogen peroxide which are produced during phagocytosis would likewise be low, another factor reducing the risk of lens damage.

Pautler & Ennis (1984) reported that pink-eyed RCS dystrophics reared in darkness had less retinal degeneration when fed a purified diet (similar to AIN-76) rather than a natural ingredient rat diet. In our rats fed a natural ingredient diet (NIH-07), dark rearing reduced retinal degeneration, but we have not studied whether feeding the AIN-76 diet would further reduce the degeneration in darkness.

SUMMARY

Light, darkness, and nutrition are the principal environmental factors that influenced the onset, or reduced the incidence and maturation of lens opacities in pink-eyed RCS rats with hereditary retinal degeneration.

When the rats were reared in cyclic light of 1-4 footcandles intensity, the percentage of rats having mature cataracts was 27-30% when the diet was comprised mainly of natural ingredients (including animal products such as fishmeal). The incidence of cataracts was reduced to 0-5% (p < 0.001) when the rats were fed (1) a purified diet (AIN-76); (2) diets lacking animal products such as fishmeal; or (3) some diets with 25% sunflower kernels as a supplement. Autoclaving one of the natural ingredient diets reduced the incidence of mature cataracts, suggesting that a volatile ingredient having a cataractogenic effect had been eliminated. Increasing the concentration of calcium in the AIN-76 diet did not significantly increase the percentage of cataracts, but substituting fishmeal for casein in the AIN diet increased the incidence from 2% to 20-38% (two batches of fishmeal). The dietary cataractogenic factors affect pink-eyed RCS dystrophic rats, but not normal rats. Diets supplemented with sunflower kernels contained higher concentrations of vitamin E and selenium, as well as moderately high concentrations of zinc and copper. These four nutrients confer antioxidant properties upon the diet. When the rats were reared in constant darkness, the rate of retinal degeneration was reduced and initiation of posterior subcapsular opacities was prevented. The concentration of macrophages in the cortex of the vitreous was greatly reduced in dark reared dystrophics, as compared with the rats reared in cyclic light. The reduced incidence of lens opacities in dystrophic pink-eyed rats reared in the dark or fed a purified diet or diets with increased antioxidant nutrients is consistent with our hypothesis that the opacities are secondary to the retinal degeneration and are initiated by toxic lipid peroxidation products from degenerating rod photoreceptor cells, and by toxic oxygen species and free radicals from macrophage activity. In dark reared rats, rhodopsin is not bleached, little retinaldehyde is set free and light is not present to interact with this sensitizer to generate singlet oxygen, which is one of the powerful oxygen species involved in peroxidation of polyunsaturated fatty acids. Consequently, fewer toxic aldehyde lipid degeneration products may be formed to harm the lens.

REFERENCES

American Institute of Nutrition (AIN). 1977. Ad Hoc Committee on Standards for Nutritional Studies. Report of the committee. J. Nutr. 107:1340-1348.

Anderson, R.E. and L.D. Andrews. 1982. Biochemistry of retinal photoreceptor membranes in vertebrates and invertebrates. In "Visual Cells in Evolution," ED., Raven Press, New York, pp. 1-22.

Anderson, R.E., L.M. Rapp and R.D. Wiegand. 1984. Lipid peroxidation and retinal degeneration. Curr. Eye Res. 3:223-227.

Bailey, J.T.J. 1959. "Statistical Methods in Biology," John Wiley and Sons, New York.

Balazs, E.A. and J.L. Denlinger. 1984. The vitreus. In "The Eye," Vol. 1A, Vegetative Physiology and Biochemistry, H. Davson, ed. Academic Press, 3rd ed., pp. 545-589.

Bellus, D. 1978. Quenchers of singlet oxygen--A critical review. In "Singlet Oxygen," Ed., John Wiley & Sons, pp. 61-110.

Bourne, M.C., D.A. Campbell and M. Pyke. 1938a. Cataract associated with an hereditary retinal lesion in rats. Brit. J. Ophthalmol. 22:608-613.

Bourne, M.C., D.A. Campbell and K. Tansley. 1938b. Hereditary degeneration of the rat retina. Brit. J. Ophthalmol. 22:613-623.

Burton, G.W. and K.U. Ingold. 1984. B-Carotene: an unusual type of lipid antioxidant. Science 224:569-573.

Caldwell, R.B. and B.J. McLaughlin. 1983. Permeability of retinal pigment epithelial cell junctions in the dystrophic rat retina. Exp. Eye Res. 36:415-427.

Delmelle, M. 1977. Retinal damage by light: possible implication of singlet oxygen. Biophys. Struct. Mech. 3:195-198.

Delmelle, M. 1979. Possible implication of photooxidation reactions in retinal photo-damage. Photochem. Photobiol. 29:713-716.

Delmelle, M., W.K. Noell and D.T. Organisciak. 1975. Hereditary retinal dystrophy in the rat: Rhodopsin, retinol, vitamin A deficiency. Exp. Eye Res. 21:369-380.

Dilley, R.A. and D.G. McConnell. 1970. Alpha-tocopherol in the retinal outer segment of bovine eyes. J. Mem. Biol. 2:317-323.

Dowling, J.E. and R.L. Sidman. 1962. Inherited retinal dystrophy in the rat. J. Cell Biol. 14:73-109.

Essner E., R.M. Pino and R.A. Griewski. 1979. Permeability of retinal capillaries in rats with inherited retinal degeneration. Invest. Ophthalmol. Vis. Sci. 18:859-863.

Farnsworth, C.C. and E. A. Dratz. 1976. Oxidative damage of retinal rod outer segment membranes and the role of vitamim E. Biochim. Biophys. Acta 443:556-570.

Fridovich, I. 1976. Oxygen radicals, hydrogen peroxide, and oxygen toxicity. In "Free Radicals in Biology," Vol. I, Ed. by W.A. Pryor, Acad. Press, New York, pp. 239-277.

Gerstein, D.D. and D.R. Dantzker. 1969. Retinal vascular changes in hereditary visual cell degeneration. Arch. Ophthal. 81: 99-105.

Goldman, A.I. and P.J. O'Brien. 1978. Phagocytosis in the retinal pigment epithelium of the RCS rat. Science 201:1023-25.

Goosey, J.D., W.M. Tuan and C.A. Garcia. 1984. A lipid peroxidative mechanism for posterior subcapsular cataract formation in the rabbit: A possible model for cataract formation in tapetoretinal diseases. Invest. Ophthalmol. Vis. Sci. 25:608-612.

Hayes, K.C. 1974. Retinal degeneration in monkeys induced by deficiencies of vitamin E or A. Invest. Ophthalmol. 13:499-510.

Heckenlively, J. 1982. The frequency of posterior subcapsular cataract in the hereditary retinal degenerations. Am. J. Ophthalmol. 93:733738.

Hess, H.H., D.A. Newsome, J.J. Knapka and G.E. Westney. 1981. Effects of sunflower seed supplements on reproduction and growth of RCS rats with hereditary retinal degeneration. Lab. Anim. Sci. 31:482-488.

Hess, H.H., D.A. Newsome, J.J. Knapka and G.E. Westney. 1982. Slitlamp assessment of age of onset and incidence of cataracts in pink-eyed, tan-hooded retinal dystrophic rats. Curr. Eye Res. 2:265-269.

Hess, H.H., J.J. Knapka, D.A. Newsome, I.V. Westney and L. Wartofsky. 1985. Dietary prevention of cataracts in the pink-eyed RCS rat. Lab. Animal Sci. 35:47-53.

Kagan, V.E., A.A. Shvedova, K.N. Novikov and Y.P. Koslov. 1973. Light-induced free radical oxidation of membrane lipids in photoreceptors of frog retina. Biochim. Biophys. Acta 330:76-79.

Kaiser-Kupfer, M., T. Kuwabara, S. Uga, K. Takki and D. Valle. 1983. Cataract in gyrate atrophy: clinical and morphological studies. Invest. Ophthalmol. Vis. Sci. 24:432-436.

Katz, M.L., K.R. Parker, G.J. Handelman, T.L. Bramel and E.A. Dratz. 1982. Effects of antioxidant nutrient deficiency on the retinal pigment epithelium of albino rats: a light and electron microscopic study. Exp. Eye. Res. 34:339-369.

Knapka, J.J., K.P. Smith and F.J. Judge. 1974. Effect of open and closed formula rations on the performance of three strains of laboratory mice. Lab. Anim. Sci. 24:480-487.

Krasnovsky, A.A., Jr. and V.E. Kagan. 1979. Photosensitization and quenching of singlet oxygen by pigments and lipids of photoreceptor cells of the retina. FEBS Lett. 108:152-154.

Kuwabara, T. and R.A. Gorn. 1968. Retinal damage by visible light: an electron microscopic study. Arch Ophthalmol. 79:69-78.

Lane, B.C. 1984. Low-level fish methylmercury as a risk factor in human cataractogenesis. Invest. Ophthalmol. Vis. Sci. 25 (Suppl.):134.

LaVail, M.M. 1979. The retinal pigment epithilium in mice and rats with inherited retinal degeneration. In "The Retinal Pigment Epithelium." K.M. Zinn & M.F. Marmor, eds. Cambridge, MA, Harvard Univ. Press, pp. 357-380.

LaVail, M.M. and B.-A. Battelle. 1975. Influence of eye pigmentation and light deprivation on inherited retinal dystrophy in the rat. Exp. Eye Res. 21:167-192.

LaVail, M.M., R.L. Sidman and C.O. Gerhardt. 1975. Congenic strains of RCS rats with inherited retinal dystrophy. J. Hered. 66:242-244.

Mullen R.J. and M.M. LaVail. 1976. Inherited retinal dystrophy: Primary defect in pigment epithelium determined with experimental rat chimeras. Science 192:799-801.

Nathan, C.F., H.W. Murray and Z.A. Cohn. 1980. Current Concepts-- The macrophage as an effector cell. New Engl. J. Med. 303:622-626.

National Research Council (NRC). 1978. Committee on Animal Nutrition, Agricultural Board. Nutrient requirements of laboratory animals, 3 rd rev.ed. National Academy of Sciences, Washington, D.C.

Noell, W.K., V.S. Walker, B.S. Kang and S. Berman. 1966. Retinal damage by light in rats. Invest. Ophthalmol. 5:450-473.

O'Keefe, T.L., H.H. Hess, T. Kuwabara and J.J. Knapka. 1985. Prevention of cataracts in the pink-eyed Royal College of Surgeons (RCS) rats by dark rearing. Invest. Ophthalmol. Vis. Sci. 26 (Suppl.):329.

Pautler, E.L. and S.R. Ennis. 1984. The effect of diet on inherited retinal dystrophy in the rat. Curr. Eye Res. 3:1221-1224.

Robison, W.G., Jr., T. Kuwabara and J.G. Bieri. 1979. Vitamin E deficiency and the retina: Photoreceptor and pigment epithelial changes. Invest. Ophthalmol. Vis. Sci. 181:683-690.

Wiegand, R.D., N.M. Giusto, L.M. Rapp and R.E. Anderson. 1983. Evidence for rod outer segment lipid peroxidation following constant illumination of the rat retina. Invest. Ophthalmol. Vis. Sci. 24:1433-1435.

Witting, L.A. 1980. Vitamin E and lipid antioxidants in free radical initiated reactions. In "Free Radicals in Biology", Vol. 4. Ed. by W.a. Pryor, Acad. Press, New York, pp. 295-319.

Zigler, J.S., Jr., R.S. Bodaness, I. Gery and J.H. Kinoshita. 1983a. Effect of lipid peroxidation products on the rat lens in organ culture: A possible mechanism of cataract initiation in retinal degenerative disease. Arch. Biochem. Biophys. 225:149-156.

Zigler, J.S., Jr., I. Gery and J.H. Kinoshita. 1983b. Macrophage mediated damage to rat lenses in culture: A possible model for uveitis-associated cataract. Invest. Ophthalmol. Vis. Sci. 24:651-654.

Zigler, J.S., Jr. and H.H. Hess. 1985. Cataracts in the Royal College of Surgeons rat: Evidence for initiation by lipid peroxidation products. Exp. Eye Res. 41:67-76.

Current Status of Vitamin E in Retinopathy of Prematurity

Neil Finer

Retrolental fibroplasia (RLF) was first described in 1942 by Terry (1942), as a condition which developed after birth initially thought to be due to the persistence of the entire vascular structure of the fetal vitreous. The term "retrolental fibroplasia" was apparently coined by a Boston opthalmologist, Doctor H. Messenger (Silverman, 1980) and by the 1950's RLF was thought to be the commonest cause of infant blindness. Following Terry's original description of RLF this condition began to be reported with increasing frequency occurring in approximately 12% of infants of less than 1400 gm in the late 1940s (Terry, 1945). The temporal changes in the retina of premature infants with RLF was first detailed by Owens & Owens of the Wilmer Eye Institute at Johns Hopkins Hospital (Owens & Owens) who described the first abnormalities as the dilation of veins, following which the arteries became dilated and tortuous. Subsequently new vessel formation was observed and was followed by retinal edema and hemorrhages. Pathologic studies demonstrated that the primary lesion was a disordered capillary endothelial proliferation within the retina which was associated with pre-retinal and vitreous hemorrhages, and fibrotic changes (Friedenwald, 1951). I have retained the term RLF as opposed to the the more recently accepted retinopathy of prematurity (ROP) because RLF was the term used in the great majority of cited papers. RLF more accurately refers to the cicatricial grades of the disorder whereas ROP has a broader connotation.

OXYGEN AND RLF

While numerous theories have been proposed to explain this disorder, the "oxygen hypothesis" has been the most popular and enduring. Thus it was postulated that changes within the retina occur secondary to high retinal oxygen concentrations in association with elevated arterial oxygen levels. Liberal oxygen administration to the low birthweight premature infant was thought to be a major factor in the development of this disorder and the clinical observations by Campbell (1951), confirmed by Patz (1952), were further supported by animal studies (Gyllensten, 1952). The studies by Ashton et al., (1954) revealed that exposure of kittens to a high oxygen environment led to constriction and subsequent obliteration of the retinal vasculature with a resultant regrowth of new vessels in a highly disorganized fashion. It was thus postulated that a hyperoxic environment led to tissue hypoxia as a result of intense vasoconstriction, a curious clinical paradox!

In an effort to confirm these observations, a cooperative study was organized under the chairmanship of Dr. V. Everett Kinsey, begun in July, 1953 and continued through until June 30, 1954. This study compared liberal vs. restricted oxygen administration for infants less than 1500 gm and the results confirmed that the incidence of RLF was significantly higher in the infants who received more than 50% oxygen (Kinsey, 1956). This study antedated the availability of arterial blood gas measurements, consequently, there were no conclusions regarding the level of arterial oxygen which was associated with retinal damage. A second prospective cooperative study was organized 15 years later by Dr. Kinsey with 5 collaborating institutions to examine the relationship between the measured arterial oxygen level (P_aO_2) and RLF. This study demonstrated that the occurrence of RLF was unrelated to any specific level of P_aO_2 as determined by intermittent blood sampling, and was more frequent among the smallest infants with the longest exposure to oxygen (Kinsey, 1977). While it would appear from the previous discussion that there has been general acceptance of the "oxygen hypothesis", this is far from the truth. There have been numerous reports of RLF in infants who never received any oxygen, and in infants with cyanotic congenital heart disease who, by the nature of their malformation, could not have had elevated arterial or retinal oxygen levels. As of 1985 the role of oxygen in the pathophysiology of RLF is still unclear (Lucy, 1984). It would appear that RLF may occur in response to any perturbation of retinal homeostasis which most likely occurs as a result of alterations in retinal blood flow and oxygen delivery. Even though there may be some doubt as to the actual etiology of RLF, almost all recent series in the literature have reported that RLF occurs with the greatest frequency and severity in the most immature infants with the longest oxygen exposure.

CURRENT INCIDENCE

The severe forms of retinopathy, the so-called cicatricial stages which indicate scarring of the retina, are seen almost exclusively in infants of less than 1500 gm birthweight with an incidence of resultant blindness in this group of between 1.8-4% (Phelps, 1981; Purohit, 1985). Five to 11% of infants less than 1000 gm are left blinded from this condition. Thus, using 1979 estimates, Phelps estimated that there would be approximately 500 blinded surviving infants with birthweights of less than 1000 gm per year in the USA (Phelps, 1981). Results regarding the incidence of RLF from the recent collaborative trial on patent ductus arteriosus (PDA) conducted between April 1979 to April 1981 revealed an overall incidence of acute RLF as can be seen in Figure 1. More advanced disease was observed up to 0-4.6% of infants from the various collaborative centres, but actual details of these grades were not provided (Purohit, 1985). However, the incidence figures from this trial probably represent the most recently available and accurate estimates by birthweight of the occurrence of this disorder in a large population of premature infants. These results have been compared to those of Reisner et al., (1985) who recently reported the incidence and severity of RLF for a large cohort of premature infants born in Tel Aviv, Israel between 1977 and 1983. In this series, no surviving infant had serious eye sequelae and early aggressive cryotherapy was the rule. We have continued to review our own incidence of RLF and over the past 2 years have noted an incidence of 60% active RLF in our 10 survivors of less than 750 gm, one of whom is blind while the other infants have no ocular morbidity on follow-up. Twenty-three percent of our 31 survivors between 751-1000 gm had RLF and no infant had any form of ocular morbidity on follow-up, for an overall rate of ocular morbidity of

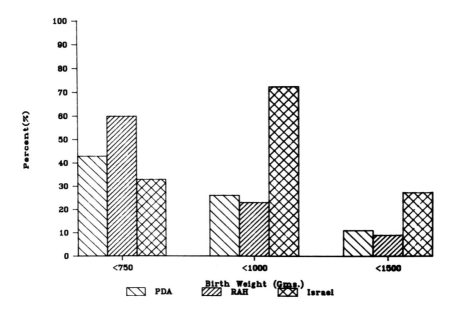

Figure 1. Incidence of acute RLF (retinopathy of prematurity) as reported in the collaborative trial on PDAs, (Purohit et al., 1985) from Israel, (Reiser et al., 1985) and from the Royal Alexandra Hospital from 1983 to 1985.

2.4% for infants of less than 1000 gm (Figure 1).

Keith and Kitchen (1984) from the Royal Womens Hospital in Victoria, Australia reported their experience from 1977 to 1982 involving 88 infants of less than 1000 gm who were followed to at least the age of 2 years. Forty-four percent of this group had evidence of RLF, of whom 17% were mild and 27% severe. Of the 88 children who were subsequently followed, 17 developed evidence of cicatricial scarring for an incidence of cicatricial disease of 19% in this population. These authors indicated that some of the eye morbidity seen with this disease, including myopia, squint, and astigmatism may not become apparent until the age of 2 years. Similar to the observations of Reisner et al, Keith and Kitchen reported that no child was blind because of RLF.

The reported incidence of RLF will obviously depend on the methods of ascertainment and, as pointed out by Keith and Kitchen, use of a speculum and indentation of the eyeball increases the detection rate and this may explain the wide variation in the reported incidence rates of RLF in the literature. Thus, while some have indicated that there has been a significant rise in the occurrence of RLF over the past decade and a resultant increase in significant morbidity (Phelps, 1981), these more recent studies have not confirmed these observations.

VITAMIN E

Vitamin E, first described in 1922 (Evans, 1922), is a fat soluble naturally occurring substance which acts as a biologic lipid antioxidant by scavenging free radical intermediates produced during oxidation of unsaturated fatty acids. A number of other potential actions have been ascribed to vitamin E which include a potential effect upon the architecture of membranes (Diplock, 1983).

In veterinary medicine, deficiency of vitamin E has been associated with a wide spectrum of disorders in a variety of animals including the pig, horse, chick, and trout. Recognized clinical deficiency states include a hemolytic anemia in premature low birthweight infants between 4-7 week of age in association with low serum tocopherol levels. This entity was first described in 1967 by Oski & Barness (1967) during a period when infant formula contained excessive amounts of linoleic acid relative to the amount of available vitamin E. Newer formulations have improved vitamin E/linoleic acid ratios resulting in a marked decrease in this disorder. Chronic vitamin E deficiency in rats, monkeys, and humans is characterized by systematic degeneration of myelinated sensory axons in the spinal cord and peripheral nerves, and has been described in children with prolonged fat malabsorption (Rosenblum, 1981; Nelson, 1983). The neurologic abnormalities in such children can be reversed or prevented with vitamin E (Alarez, 1985). Apart from deficiency states, there are few, if any, defined clinical indications for this agent.

Numerous animal research studies have demonstrated that vitamin E can protect various cells from oxygen-induced damage, and because of clinical concerns regarding oxygen toxicity to the lung this organ has received the most attention. Wender et al., (1981) reported that vitamin E prevented early lung epithelial injury in oxygen-exposed newborn rabbits. These observations were consistent with initial observations by Ehrenkranz et al., (1978) who reported that neonates with respiratory distress syndrome (RDS) who received intramuscular (IM) vitamin E had a significantly lowered incidence of bronchopulmonary dysplasia, (BPD) a form of chronic lung disease seen in up to 35% of neonates who require mechanical ventilation and thought to occur as a result of oxygen toxicity in addition to other factors. Concerns were expressed regarding the comparability of the two groups in this original study and the study design was questioned (Northway, 1978). A subsequent study by the same group (Ehrenkranz et al., 1979) (a prospectively randomized, double-blind evaluation of IM vitamin E in which 36 neonates with RDS were evaluated) failed to establish any significant effect of vitamin E. These results were confirmed subsequently in a study by Saldanha et al., (1982) and at least one subsequent study in the newborn lamb showed that pharmacologic vitamin E administration to these animals did not protect them from oxygen-induced lung injury (Hansen, 1982).

The preceding discussion of vitamin E and BPD is in many ways a parallel to the history of the use of vitamin E in RLF, the latter being a much more devious tale spanning a significantly longer interval of time.

VITAMIN E AND RLF

Only 7 years after the original description of the disease, Owens & Owens (1949), who themselves had made very important contributions to the definition of this disorder suggested that vitamin E deficiency may be a significant factor in the development of RLF. They noted that vitamin E was the only fat soluble vitamin not routinely administered to newborns, and drew a parallel between the central nervous system lesions produced in vitamin E deficient animals and RLF. They postulated that the administration of vitamin E would reduce the incidence of RLF. In their initial study they showed that RLF developed in none of their 11 vitamin E treated infants compared with 5 of the 15 of their control infants, and they established that serum tocopherol levels in the blood of the vitamin E treated infants were significantly elevated demonstrating that vitamin E was well absorbed (Owens & Owens, 1949). In view of these very encouraging preliminary results, the controlled trial was abandoned and all subsequent low birthweight infants in their nurseries were given supplemental vitamin E. The overall incidence of RLF was 4.4% in the 101 premature infants of less than 1360 gm who received vitamin E vs. 21.8% in the controls. Others were unable to confirm these encouraging results (Laupus, 1951; Reese, 1951) and there were no further trials evaluating the use of vitamin E in RLF until 1974. This no doubt reflected the intense interest being placed on the etiologic role of oxygen in this disorder. Following the completion of the first cooperative trial by Kinsey et al., (Kinsey,1956) severe restrictions were placed on the use of oxygen in nurseries which wes associated with a reduction in the incidence of RLF (albeit with an associated increase in neonatal morality and morbidity (Silverman, 1980).

The next documented experience regarding vitamin E and RLF was that of Johnson et al., who in 1974 reported that maintenance of a serum tocopherol level of 1.5 mg/dl was associated with a reduction in the incidence of RLF (37.5% of the vitamin E treated infants compared with 70.5% of their infants on placebo). Their numbers were small, however, (33) and this reduction failed to reach significance (p=.059). In this study they also noted a tendency toward more healing and regression in the infants who had received vitamin E (Johnson, 1974).

REVIEW OF CURRENT TRIALS

In the following section I will attempt to review the published studies which have appeared since the report by Johnsonet al., (Johnson, 1974) regarding vitamin E and RLF in the neonate. The first of these studies was published by Hittner et al., (1981). In this prospective, randomized, controlled, double-blind trial, stratified into 250 gm weight groups, 150 infants of 1500 gm or less who required supplemental oxygen administration for respiratory distress, seen within the first 24 hr of life, were studied. The treatment group received 100mg/kg of oral vitamin E compared with the control group who received 5 mg/kg/dy. Forty eight of the infants enrolled died prior to the first 4 weeks of life and were not included in the analysis, and 1 further infant was excluded because he was unable to be fed for the first 10 days of life. The efficacy of vitamin E was analyzed by comparing the most advanced grade of active RLF which developed before spontaneous regression or surgical intervention using the McCormick classification (1977). In this study, when infants reached Grade 3 which included vitreal neovascularization with increased dilation and tortuosity, surgical intervention with cryotherapy and scleral buckling was performed

(McPherson, 1979). The results of this study revealed a significant reduction in RLF equal to or greater than Grade 2 with no overall reduction in the incidence of RLF (Table I). Multivariate analysis revealed that vitamin E significantly reduced the severity of RLF (p=.012). The average plasma level of vitamin E in the infants receiving 100 mg/kg between day 7-35 was 1.2 mg/100 ml compared with .6 mg/100 ml for the controls.

This group continued to investigate the role of vitamin E and RLF, and in subsequent studies (Hittner, 1982; Hittner, 1983; Kretzer, 1982) confirmed their original observations. In 1983 they compared the outcome of 100 infants of less than 1500 gm who received 100 mg/kg of oral vitamin E daily to the 75 control infants in their first study. (Hittner, 1983) vitamin E did not produce a significant reduction in the incidence or severity of RLF by univariate analysis. Using multivariate analysis vitamin E was shown to significantly reduce the severity of RLF (p=.003), even when the 6 infants who died after 10 week were excluded. These authors argued that multivariate analysis is more appropriate than univariate analysis because of the multiplicity of risk factors which are related to the development of RLF.

In 1984 Hittner et al., (1984) reported the results of a trial which evaluated the efficacy of additional early IM vitamin E to infants receiving oral therapy. This study involved 168 infants of less than 1500 gm who required supplemental oxygen for respiratory distress, of whom 33 died before the age of 10 week. In this study, 68.7% (46 of 67 control infants) developed acute RLF and 70.6% of the vitamin E group (48 of 68) developed RLF, with 1 control and 3 treated infants developing Grade 3 RLF. All of the Grade 3 RLF occurred in infants of less than 1000 gm and the incidence of all grades of RLF for infants under 1000 gm was 86% in the controls vs 78% in the treated infants. In this study, the treatment dose of vitamin E did not decrease the occurrence of severe RLF, indicating that the provision of additional IM vitamin E to infants already receiving oral vitamin E provided no additional benefits in terms of reducing the severity of RLF (Table I). Hittner et al., (1984) also noted a significantly decreased incidence of intraventricular hemorrhage (IVH) in infants who received supplemental IM vitamin E, consistent with the observations of Chiswick et al., (1983), and these findings were the subject of a separate report (Speer et al., 1984). These observations are of interest because of the potential

TABLE I

Author	Ref #	Route	Total Control	Vit E	Total Control	Vit E
Hittner, et al	(1981)	Controlled oral	33/51	32/50	5/51	0/50
Hittner, et al	(1984)	Oral vs. IM	46/67	48/68	1/67	3/68
Puklin, et al	(1982)	IM	9/37	8/37	1/37	3/37
Milner, et al	(1981)	Oral	19/114	13/111	5/114	3/111
Watts, et al	(1985)	Oral	23/73	28/75		
Finer, et al	(1982)	Controlled IM	12/51	9/48	4/51	0/48
Schaffer, et al	(1985)	IV	109/216	95/208	9/216	3/208
Phelps, et al	(1985)	IV	28/99	25/97	8/99	11/97
			$p > .05$		$p > .05$	

association between RLF and IVH as pointed out by the group from
Baylor (Procianoy, 1981.)

Dr. Frank Kretzer, a member of the Cullen Eye Institute at the
Baylor College of Medicine and a collaborator in these studies, had
begun a detailed ultrastructural histologic examination of the eyes
of premature infants who died in the Texas Children's Hospital
Neonatal Intensive Care Unit in Houston. In the first report of
these results, eyes from 11 infants between 25-29 week gestational
age were studied. These eyes were examined by light and electron
microscopy. The hypothesis put forward by Kretzer et al., was that
the linkage of spindle cells by gap junctions inhibits the migration
of new vessels and stimulates the proliferation of spindle cells
which produces the disorganized new vessel formation in RLF.
Spindle cells are mesenchymal cells which arise from the division of
cells in the hyaloid artery and subsequently invade the nerve fiber
layer to form the vanguard of the vasoformative elements of the eye.
These gap junctions are thought to be specialized areas of adjacent
plasma membranes and involved in the communication between cells.
Kretzer et al., (1982) postulated that an increase in the gap
junction area may be the cellular response to hyperoxia, perhaps
mediated by free oxygen radicals which would induce lipid
peroxidation and alter membrane properties. They hypothesized that
sufficient levels of vitamin E would suppress such an increase in
gap junctions, perhaps through its antioxidant effect. They noted
that the spindle cells in the retinas of infants who received oxygen
for respiratory distress showed an increase in gap junction area
which was suppressed by vitamin E therapy. They suggested that such
suppression required high levels of vitamin E which may take up to
seven days when infants are treated orally, accounting for some
treatment failures using early oral therapy. They also noted a

failure to suppress gap junctions in very immature infants of less than 27 week gestation. These studies were used to support the authors' findings that vitamin E reduced the severity of RLF in infants of greater than 27 week gestation. In a follow-up study (Hittner, 1982) this group extended their histologic studies by reporting on the histologic examination of eyes from 63 infants and confirmed that treatment levels of vitamin E failed to alter increases in gap junctions in infants of less than 27 week, whereas, such treatment in more mature infants prevented such increases.

Others have taken exception to the notion that gap junctions play a pivotal role in the development of RLF, and Gole et al., (1985) have pointed out the difficulty in both recognizing and interpreting gap junctions on post mortem specimens. This group has also emphasized the fact that the relationship of gap junctions to RLF remains speculative and requires further research. If vitamin E acts to inhibit the formation of gap junctions and thus prevent the activation of the spindle cell, then why is the overall incidence of RLF not decreased by such therapy? This question was originally posed by Ehrenkranz and Puklin (1982) and has not yet been adequately addressed.

Pucklin et al., (1982) studied 100 neonates admitted with a diagnosis of RDS and seen within the first 24 hr of life who were randomized to receive either IM vitamin E, or a placebo. The infants received 20 mg/kg on admission and then 24,48,168 hr later with additional doses being given twice weekly for as long as the infant remained in an oxygen-rich environment and was unable to tolerate oral feedings and vitamin supplements. In this study, all infants received oral vitamin E preparation, Aquasol E, (USV Pharmaceuticals) in a dose of 50 units for infants less than 1000 gm, and 25 units for infants over 1000 gm given as soon as the infants regularly tolerated feedings. After removing the early deaths, there were 37 infants in each group who had adequate eye examinations and of these infants, 9 of the treated vs. 8 of the controls had RLF of Stages I & II described by the authors. Thus, there was no significant difference in either the incidence or the severity of RLF in spite of the marked increase in serum tocopherol levels observed in the treated infants (Table I). In this study the serum tocopherol of the placebo treated group was 1.11 unit by the second week of life, a value which is in the normal range for adults, no doubt reflecting the routine administration of oral vitamin E to all infants on established oral feedings. No longterm follow-up was given regarding the final stage of the eye disease or the longterm visual sequelae, a shortcoming of all the previously quoted recent trials.

Milner et al., (1981) reported on the results of a prospective, controlled, randomized trial to evaluate the effects of vitamin E on RLF, BPD, and anemia of prematurity. The treated group received 25 units of d-1 alpha tocopherol daily for 6 week of life or until the baby was well enough to be discharged back to a referral unit. Control infants received an identical vehicle, thus the study was blinded. The supplemented group had serum tocopherol levels of 2.86 mg% at 1 week and 3.16 mg% at 3 week of age. Thirteen of 110 infants who received vitamin E compared with 19 of 114 in the control group developed RLF, and these differences were not significant (Table I). Using multivariate analysis vitamin E therapy did appear to significantly reduce severe RLF in this trial, but the total variation explained by the stepwise multiple regression was only 10.6%. This trial continued until 266 infants were enrolled and 148 survivors with retinal examinations were available for analysis (Watts, 1985), and RLF was diagnosed in 28 of the 75 vitamin E treated infants compared with 23 of the 73 controls (Table I). There was no significant difference in the incidence or severity of RLF between the two groups.

In 1982 we published the results of a controlled prospective evaluation of IM vitamin E on RLF (Finer et al., 1982). We had been stimulated by the previous results of Johnson to evaluate IM vitamin E and used a preparation of d-1 alpha tocopherol (provided by Hoffman-La Roche) in a dosage schedule very similar to that used by Johnson et al., (1982). Eye exams began at 4 weeks of age and subsequent intervals were determined by the ophthalmologists. All infants were seen at discharge and any infant with an ocular abnormality was followed until complete regression or cessation of the eye disease. Of 126 infants enrolled, there were 27 infants who died before one month of age, and thus were not evaluated for the presence or absence of RLF. Of the remaining infants, 48 received vitamin E and 51 were controls, and the groups were otherwise comparable. Results revealed a significant increase in vitamin E levels in the treated group to levels that were almost 4 times as high as that observed in Hittner's group, (Day 3 the level was 2.9 mg/dl in the treated group). In this study, any infant with evidence of RLF was continued on vitamin E until the eye disease was judged to be quiescent by the ophthalmologist. There was no significant difference in the incidence or severity of RLF between the treated or control groups using univariate analysis (Table I). As with all other studies regarding RLF, infants with this disease were significantly younger and had longer oxygen exposure than infants without this disorder. Using multiple linear regression analysis, vitamin E therapy was shown to significantly reduce the severity of RLF (p <.01).

Following completion of the IM trial we began to administer oral vitamin E, initially using the commercially available preparation (Aquasol-E, USV) at 100 mg/kg. We analyzed the outcomes of 92 infants who received early vitamin E, 74 who received no vitamin E, and 8 who received vitamin E but only after 39 hr of age (Finer et al., 1983). We found a significant increase of all forms of cicatricial RLF in infants who had not received vitamin E, and a significant increase in the occurrence of Grade 3 or greater cicatricial RLF in such infants (p<.017). Multivariate analysis, comparing infants with cicatricial RLF at eye follow-up with the remaining group revealed only two significant predictors of the severity of the visual sequelae, and they were the early administration of vitamin E (p<.0001), and the number of days of oxygen exposure (p=.0002).

Lois Johnson and her collaborators have certainly exhibited the most enduring interest in vitamin E and RLF. Her studies began in 1968 and continue to the present (Johnson et al., 1985). Their previous study in 1974 (Johnson et al., 1974) rekindled significant interest in the issue of vitamin E and RLF. This group produced a modification of the standard classification of RLF and introduced a concept of "plus" grades which were characterized by circumferential retinal involvement. Between 1974-1976 this group compared IM alpha tocopherol acetate with alpha tocopherol alcohol in a design which involved different temporal treatment periods. From 1976 to mid 1978, infants with severe acute RLF were prospectively identified and treated with high doses of IM and oral vitamin E to increase the serum tocopherol level to 5-6 mg/dl. They maintained these levels until progression of the disease had stopped and retinopathy had begun to regress, following which the doses were decreased to maintain levels at 3-4 mg/dl for a further month. A level of 2.5 mg/dl was maintained until one year of life. During the four years when vitamin E was being assessed in the nursery, there were more sequelae evident in the control infants than the treated infants, lending some support to the role of the vitamin E in this disorder. These authors also observed a significant reduction in sequelae in their control infants and attributed this to the change in infant formulas from the 1960's during which formulas contained high amounts of iron and polyunsaturated fatty acids (PUFA) and low levels of vitamin E. With the reduction in iron content and the improved ratio of vitamin E to PUFA throughout the 1970's in infant formulas, these authors postulated that infants became more vitamin E sufficient. They were able to show a significant increase in the serum tocopherol levels in unsupplemented infants from 1972 to 1978 such that by 1978 infants who had not received any supplemental vitamin E, but who had been able to take early oral feeds, had vitamin E levels by 2 weeks of age within the normal adult range.

Ten infants were treated for Grade 3+ active RLF with large doses of parenteral vitamin E between 1976 and 1978. These infants were compared with 14 infants from the same period with apparently similarly severe RLF, but who were not treated with vitamin E. The authors compared these two groups and noted a significantly improved visual outcome in the vitamin E group (40% vs. 14% with good outcomes). It should be remembered, however, that this was not a prospective, randomized, controlled trial, and that these infants who were treated were selected for therapy because of their eye disease, and the infants who did not receive therapy were placed in this class either because they had a major retinal detachment or because treatment had not been requested.

This group then began a prospective, multiinstitutional, double-blinded trial comparing parenteral vitamin E with a placebo. This study was designed to assess the effectiveness of high doses of vitamin E, and thus, in the treatment group the target level of serum tocopherol was 5mg/dl, chosen because of their previous experience that such levels improved visual outcome in infants with established severe acute RLF. Nine hundred and fourteen infants were enrolled between January 1979 and May 1981, of whom 424 were 1500 gms or less at birth and had sufficient information for detailed analysis regarding acute RLF. vitamin E did not reduce the overall incidence of acute RLF (109/216 controls vs. 95/208 vitamin E) nor did it significantly affect the severity, although there were fewer infants with Grade 3 "plus" disease in the treated group

(9/216 vs. 3/208, p>.05) (Table I). In this trial all infants who reached Grade 3+ retinopathy were removed from the study and placed on vitamin E to achieve the previously described target serum levels. Three hundred and twenty-eight babies were followed to one year and vitamin E did not reduce the incidence of cicatricial RLF. There were 70 infants whose birthweights were less than 1000 gm seen at one year making this the largest trial reported for this high risk group, and overall the incidence of cicatricial RLF was 16% for these infants. There were fewer infants with severe eye sequelae in the vitamin E group, (8/154 vs. 3/163, p>.05) but this difference did not reach significance (Schaffer et al. , 1985). In an addendum to this trial, the authors have indicated that reanalysis of their data has indicated that vitamin E was associated with a significant reduction in the severity of RLF.

In addition to these previously published trials, Curran & Cantolino (1978), presented preliminary data on 20 infants, 10 of whom had received injectable vitamin E and all of whom had received fluorescein angiography. They noted a reduction in the incidence of RLF in infants who received vitamin E, (20% vs. 60% in their controls). In another abstract, McClung et al., (1980) reported on 28 infants, 14 of whom received vitamin E, and they were unable to find any benefit associated with the development of RLF.

While some have been very enthusiastic about the role of vitamin E in RLF, others have remained skeptical, the most prominent of whom has been Dale Phelps, who in collaboration with Arthur Rosenbaum, had demonstrated a beneficial effect of vitamin E on oxygen-induced retinopathy in kittens. In a subsequent review of vitamin E in RLF, she postulated that in order to prospectively assess the benefits of vitamin E in the group of infants at sufficient risk, (infants less than 1000gm who had approximately a 25% chance of developing cicatricial RLF) a definitive study would require a total sample of 200 infants of this birthweight (Phelps, 1982). Phelps et al. (1985) recently reported the results of a prospective, controlled, doubleblinded evaluation of vitamin E and RLF. There were 287 infants of 1500 gm or less, and the treatment and control groups were well matched. The vitamin E group received IV vitamin E followed by oral vitamin E with serum tocopherol levels maintained at 3-3.5 mg/dl. There were no significant differences in mortality, sepsis, or NEC, although there apparently was an increased incidence of Grade 3 & 4 intracranial hemorrhages in infants less than 1 kg who received vitamin E. The incidence of acute RLF and Grade 3 or greater RLF was not significantly different between the two groups (Table I). Only one infant in each group developed Grade 4 cicatricial RLF and more infants who received vitamin E developed retinal hemorrhages than the control infants (Rosenbaum, 1985). In this study there were only 36 infants of less than 1000 gm (18 in each group) far less than the ideal number required as postulated by Phelps.

ANIMAL RESEARCH: VITAMIN E AND RLF

While the above review has attempted to focus on clinical trials, numerous animal studies have suggested a variety of potential benefits of vitamin E on RLF. Support for the role of vitamin E in ameliorating RLF was found in the research of Phelps & Rosenbaum (1977) who demonstrated in an initial study that vitamin E given to newborn kittens prior to oxygen exposure reduced the incidence and severity of the RLF like disease seen in these animals. In a subsequent study the same group as able to demonstrate an inhibition of intravitreal neovascularization by the administration of vitamin E following oxygen exposure using the same kitten model.(Phelps & Rosenbaum 1979a ; Phelps et al., 1979). The observation that tocopherol given after oxygen exposure reduced vitreal neovascularization suggested a different mechanism of action; that tocopherol acted to potentially suppress the formation of collagen in new vessels as had been previously described in rats (Erlich, 1972). While these studies may be pertinent to RLF in the human, there have been concerns expressed because the kitten model does not develop cicatricial changes or retinal detachment. Nevertheless, these two animal studies gave strong support to the notion that vitamin E could be beneficial in reducing either the incidence or severity of RLF, and were in support of Johnson et al's earlier observations (Johnson, 1974). More recently, Phelps re-examined the role of vitamin E in a study in which she compared vitamin E, aspirin, indomethacin, and hypercarbia in a group of kittens who were subsequently placed in a hyperoxic environment (Phelps, 1981). In this study, none of the tested drugs were able to attenuate the vasoconstriction seen following hyperoxia, suggesting vitamin E does not exert its beneficial effect by maintaining vessel patency.

Bougle et al., (1982) demonstrated that pretreatment with 100 mg of alpha tocopherol acetate in kittens who were subsequently exposed to a hyperoxic environment led to a higher level of retinal superoxide dismutase activity than in untreated animals. Superoxide dismutase is an enzyme which protects cellular metabolism by scavenging the superoxide radical, and its levels are usually increased by hyperoxia. Thus, this study suggested a potential mechanism by which vitamin E may exert its protective effect in RLF. In addition, Stuart et al., (1981) demonstrated that vitamin E increased the ability of neonatal plasma to regenerate prostacyclin, a prostaglandin which causes vasodilation and inhibits vascular thrombus formation in vessels that were artificially depleted of this substance. These results were consistent with previous observations that vascular prostacyclin formation is reversibly decreased in vitamin E-deficient animals (Okuma, 1980; Chan, 1981). Therefore, it may be postulated that by increasing the amount of available prostacyclin, vitamin E may inhibit vasoconstriction and platelet aggregation in the retina secondary to hyperoxia. In addition, vitamin E may act by decreasing platelet aggregation, and thus decrease the thrombosis which occurs subsequent to retinal vasoconstriction (Fong, 1976). It is most interesting, however, that clinical trials have not been able to demonstrate that vitamin E in pharmacologic doses decreases platelet aggregation (Gomes, 1976; Oske, 1977). The proposed mechanisms of action of vitamin E are presented in Table II.

VITAMIN E TOXICITY

There have been suggestions that vitamin E does have a significant toxicity in human newborn infants, especially when used in pharmacologic doses. Animal studies have demonstrated dose related toxic effects. Parenteral doses of 200 mg/kg for one week followed by 100 mg/kg for two weeks was associated with an 11% mortality with associated progressive lethargy, weight loss, and seizures in 5 of 44 kittens between 2-3.5 week of age. One thousand mg/kg of free vitamin E was uniformly fatal, preceded by lethargy and poor feeding (Phelps, 1981). There was evidence of hepatomegaly, but no significant difference in liver function observed in 23 kittens who received 50 mg/day of vitamin E acetate or placebo. Hepatomegaly was noted with levels of 10 mg% and over, and mortality

TABLE II
PROPOSED MECHANISMS OF VITAMIN E

Antioxidant	Non-Antioxidant
-Directly prevent oxygen damage	-Direct inhibition of vessel growth
-Prevent hypoxic induced acute oxidation	-Anti-inflammatory
-Spindle cell activation	
-Inhibit collagen cross-linkage	
- prostacyclin, platelet aggregation	
- retinal SOD	

increased as levels rose to 24 mg/dl.

In a study which evaluated the tissue levels of vitamin E in newborn rabbits, Knight & Roberts (1985) noted that the concentration of tocopherol was highest in the liver after either subcutaneous or oral administration, or IV administration of either alpha tocopherol or alpha tocopherol acetate, whereas, IV administered alpha tocopherol acetate was associated with peak concentrations in the lung. Thus, tissue deposition and possible tissue toxicity may depend on the route of administration and composition of vitamin E. These authors pointed out that the serum tocopherol levels do not necessarily reflect vitamin E tissue concentrations and that the use of such levels may lead to inappropriate assumptions of efficacy.

More recently, preliminary observations on tissue vitamin E content from human infants who received vitamin E have been reported (Roberts, 1985). This study showed that IV administered tocopherol acetate was associated with very high levels of tocopherol acetate in the liver (30-500 µg/gm), similar to the observations of the previous study by Knight & Roberts (1985). The study also showed that infants who received only nutritional amounts of tocopherol acetate (2-25 units/day) had tocopherol levels of less than 20 µg/gm in the liver, kidney, and lungs. Bhat & Dahiya (1985) reported on the levels of vitamin E in the retina, choroid, and vitreous on some 20 eyes of deceased neonates who had received 50 mg/kg/day of vitamin E. They found no correlation between the serum and retinal levels. In addition, retinal levels in the treated group were significantly higher than the untreated group, and these levels increased with gestational age confirming the lack of any relationship between serum and retinal levels of vitamin E. Interest in clinical vitamin E toxicity has recently been stimulated by a number of clinical reports. The first published report of the large prospective controlled trial of Johnson et al., (1985) (infants enrolled from January, 1979 through to April, 1981) described the occurrence of two neonatal complications namely, sepsis or infection, and necrotizing enterocolitis. This second disorder is one frequently seen in neonates, occurring in approximately 10-15% of very low birthweight infants, and usually associated with antecedent asphyxia and a postulated subsequent reduction of blood supply to the bowel resulting in bowel ischemia.

A total of 914 infants were enrolled in this study of whom 545 were less than 1500 gm. Infants in the treated group had a significantly greater incidence of sepsis after the first week of life and of necrotizing enterocolitis (NEC) after 2 weeks of life, indicating that there was association between the duration of treatment and the development of these complications. The incidence of sepsis was 10.2% in the placebo group vs. 18.5% among the vitamin E treated infants (p=.009), and the incidence of NEC was 10.2% among the placebo group vs. 15.6% among the vitamin E treated infants. This difference was not significant. The overall pattern of sepsis and NEC episodes were significantly more frequent in the treated group. When only infants who received the study medication for more than 8 days were evaluated, 9.5% of placebo infants vs. 19.4% of the vitamin E treated infants developed sepsis (p=.0039), and 9.5% of the placebo treated vs. 15.9% of the vitamin E treated infants developed NEC (p=.037).

The authors attributed these observations to the high levels of vitamin E which were thought to compromise the ability of white cells to kill ingested bacteria, perhaps related to the marked decrease in hydrogen peroxide within the polymorphonuclear cell which is associated with vitamin E therapy (Baehner, 1982). The polymorphonuclear cells of sick neonates are known to have a decreased metabolic reserve and demonstrate impaired chemotaxis and intracellular killing capacity (Wright, 1975; Anderson, 1975; Shigeoka, 1978; Laurenti, 1980), and thus vitamin E in large doses may further compromise the immune function of neonatal white cells.

In a previous study we had demonstrated an increased incidence of NEC in low birthweight neonates who had received a hyperosmolar oral preparation of vitamin E (Finer, 1984). We felt at that time that this might be due to the actual preparation and its high osmolar load. In addition, our infants who developed NEC did not have vitamin E levels in the range described by Johnson etal., (1985). However, Johnson et al., (1982) did not note any association between the osmolality of feeds and the development of NEC since both their placebo infants and the infants who received vitamin E received a preparation of similar osmolality, and in their study they felt the increased incidence of NEC was related to the high levels of vitamin E.

Of even greater concern was the initial report of 3 infants who died with clinical evidence of a coagulopathy and renal and hepatic failure following the institution of an IV vitamin E preparation known as E-ferol. The vitamin E level on Day 11 in one of these patients was extremely elevated (12.9 mg/dl) (Bodenstein, 1984). Reports describing similar illnesses appeared from a nursery in Knoxville, Tennessee in which 8 infants developed symptoms of whom 5 died (Butler, 1984; Lorch, 1985). This drug was subsequently withdrawn from the market.

It has been postulated that the toxicity of E-ferol may be related to the actual toxicity of the tocopherol itself, or to polysorbate 80, an additive, or to possible interactions between the vitamin E and other administered IV products such as hyperalimentation solutions. In addition, Balistreri & Bove (1985) have described a distinctive hepatopathy associated with E-ferol therapy in autopsy material from infants who died following E-ferol administration. They believe that this pathology is secondary to the toxic effects of E-ferol. While these effects may be related to E-ferol, a drug which was never available for use in Canada, we have seen 3 infants whose livers at post mortem had histologic findings consistent with those of Balistreri & Bove. Thus, we do not believe this lesion to be specific to E-ferol, but may in fact represent a toxicity of vitamin E in keeping with animal studies suggesting that the liver is a major target organ. Alternately, these lesions may be the result of the toxic interaction of vitamin E with other IV substances, etc. Thus, this product which was only on the market for seven months was associated with approximately 38 deaths.

In addition to the previously described toxicities, Smith & Goss (1969) reported that IM vitamin E was associated with extensive ossification in the thigh in 2 infants who received this medication. In view of the reported toxicity of vitamin E and the recently reported trial by Phelps et al., (1979) failing to show efficacy in a large prospective cohort, the Committee on the Fetus and Newborn of the American Academy of Pediatrics issued the statement that "The Committee regards the prophylactic use of pharmacologic vitamin E experimental and cannot recommend that high doses of vitamin E be given routinely to infants weighing less than 1500 gm, even if such use is limited to infants who require supplemental oxygen." (Committee on the Fetus and Newborn. 1985)

CONCLUSION

In conclusion, 37 years after the initial report by Owens & Owens indicating that vitamin E may reduce the severity of RLF, one cannot state unequivocally that vitamin E significantly alters the course of RLF. Some intensive care units because of their previous experience, including our own, continue to treat all oxygen exposed low birthweight infants with vitamin E while others have not adopted the use of vitamin E in doses any larger than those required to produce sufficiency in the neonate. Whereas it does not appear as if the absolute incidence of RLF, especially severe grades which produce cicatricialization of the retina and subsequent retinal morbidity are on the increase, the survival of the very low birthweight neonate is improving and morbidity from this disease will continue to represent significant concern.

If there is a message to be taken from all of this previous experience, it is perhaps that well intentioned, even prospectively controlled and randomized trials with insufficient numbers can only increase our confusion regarding various treatment modalities. Before any new form of therapy is accepted, it should be proven to be efficacious with an acceptable risk benefit ratio, and it would appear that this can only be done in a reasonable period of time by large collaborative trials organized by knowledgeable individuals receiving enthusiastic support. In my opinion, such a study is required to determine the ultimate role of vitamin E in RLF and the development of an International Classification of Retinopathy of Prematurity (The Committee for the Classification of Retinopathy of Prematurity. 1984) should facilitate such a trial.

"SEE HOW TODAY'S ACHIEVEMENT IS ONLY TOMORROW'S CONFUSION"

Wd Howells

REFERENCES

Alarez F, Landrieu P, Feo C, et al., 1985. Vitamin E deficiency is responsible for neurologic abnormalities in cholestatic children. J Pediatr 107:422-425.

Anderson D, Pickering L, Feigen R: 1975. Leukocyte function in normal and infected neonates. J Pediatr 85:420-425.

Ashton N, Ward B, Serpell G. 1954. Effect of oxygen on developing retinal vessels with particular reference to the problem of retrolental fibroplasia. Br J Opthalmol 38:397-432.

Baehner RL, Boxer LA,Ingraham LM, et al., 1982. The influence of vitamin E on human polymorphonuclear cell metabolism and function. Ann NY Acad Sci 393:237-250.

Balistreri WF, Bove KE. 1985. Distinctive hepatopathy in low birthweight infants in association with E-ferol. Pediatr Res 19:212A.

Bhat R, Dahiya U. 1985. Vitamin E in human eye retinal, choroidal and vitreous levels with and without treatment. Pediatr Res 19:334A.

Bodenstein CJ. 1984. Intravenous vitamin E and deaths in the intensive care unit. Pediatrics 73:733.

Bougle D, Vert P, Reichart E, et al., 1982. Retinal superoxide dismutase activity in newborn kittens exposed to normobaric hyperoxia. Effect of vitamin E. Pediatr Res 16:400-402.

Butler J, Hutchison M, Sandlin M. 1984. Deaths in preterm infants associated with intravenous vitamin E supplement. Am J Hosp Pharm 41:1514-15.

Campbell K. 1951. Intensive oxygen therapy as a possible cause of retrolental fibroplasia: A clinical approach. Med J Aust 2:48-50.

Chan AC, Leith MK. 1981. Decreased prostacyclin synthesis in vitamin E deficient rabbit aorta. Am J Clin Nutr 34:2341-2347.

Chiswick ML, Johnson M, Woodall, et al., 1983. Protective effect of vitamin E (d-1 alpha tocopherol) 36 against intraventricular hemorrhage in premature babies. Br Med J 287:81-84.

Committee on the Fetus and Newborn. 1985. Vitamin Eand the prevention of retinopathy of prematurity.Pediatrics 76:315-16.

Curran JS, Cantolino SJ. 1978. Vitamin E (injectable) administration in the prevention of retinopathy of prematurity. Evaluation with fluorescein angiography and fundus photography. Pediatr Res 12:404.

Diplock AT. 1983. The role of vitamin E in biological membranes. Biology of vitamin E. Pitman Books, London (Ciba Foundation Ehrenkranz symposium 101), p45-55.

Ehrenkranz RA,Bonta BW,Ablow RC, et al., 1978. Amelioration of bronchopulmonary dysplasia after vitamin Eadministration. A preliminary report. N Engl J Med 299:564-569.34

Ehrenkranz RA, Ablow RC, Warshaw JB. 1979. Prevention of bronchopulmonary dysplasia with vitamin Eadministration during the acute stages of respiratory distress syndrome. J Pediatr 95:873-878.

Ehrenkranz RA, Puklin JE. 1982. Letter to the editor. Opthalmology 89:988-989.

Erlich HP, Tarver H, Hunt TK. 1972. Inhibitory effects of vitamin E on collagen synthesis and wound repair. Ann Surg 175:235-240.

Evans H, Bishop KS. 1922. The existence of a hitherto unrecognized dietary factor essential to reproduction. Science, Washington, D.C., 56:620-661.

Finer NN, Schindler RF, Grant G, et al., 1982. Effect of intramuscular vitamin E on frequency and severity 37of retrolental fibroplasia. A controlled trial. The Lancet, May 15, 1987-91.

Finer NN, Schindler RL, Peters KL, et al., 1983. Vitamin E and retrolental fibroplasia: Improved visual outcome with early vitamin E. Ophthalmology 90:428-435.

Finer NN, Peters KL, Hayek Z, et al., 1984. Vitamin E and Necrotizing Enterocolitis. Pediatrics 73:387-393.

Fong JSC. 1976. Alpha tocopherol: its inhibition on human platelet aggregation. Experientia 32:639-41.

Friedenwald JS, Owens WC, Owens EU. 1951. Retrolental fibroplasia in premature infants. III. Pathology of the disease trends. Trans Am Ophthalmol Soc 49:207.

Gole GA, Skinner JM, Henderson DW, et al., 1985. Vitamin E and retinopathy of prematurity revisited. Pediatrics 75:1166-1167.

Gomes JAC, Venkatachalapathy D, Haft JI. 1976. The effect of vitamin E on platelet aggregation. Amer Heart J 99:425-429.

Gyllensten LJ, Helmstrong BE. 1952. Retrolental fibroplasia animal experiments: The effect of intermittently administered oxygen on the postnatal development of the eyes of fullterm mice. Acta Pediatr Scand 41:577-582.32

Hansen TN, Hazinski TA, Bland RD. 1982. Vitamin E does not prevent oxygen induced lung injury in newborn lambs. Pediatr Res 16:583-587.

Hittner HM, Godio LB, Rudolph AJ, et al. 1981. Retrolental fibroplasia: efficacy of vitamin E in a double 35 blind clinical study of preterm infants. N Engl J Med 305:1365-71.

Hittner HM. 1982. Response to letters to the editor. N Engl J Med 306:867.

Hittner HM, Kretzer FL. 1983. Vitamin E and retrolental fibroplasia: ultrastructural mechanism of clinical efficacy. Biology of vitamin E. Pitman Books, London (Ciba Foundation symposium 101) p 165-185.

Hittner HM, Godio LB, Speer ME, et al., 1983. Retrolental fibroplasia: further clinical evidence and ultrastructural support for efficacy of vitamin E in the preterm infant. Pediatrics 71:423-32.

Hittner HM, Speer ME, Rudolph AJ, et al., 1984. Retrolental fibroplasia and vitamin E in the preterm infant comparison of oral vs. intramuscular: oral administration. Pediatrics 73:238-249.

Johnson L, Bowen FW, Abbasi S, et al., 1985. Relationship of prolonged pharmacologic serum levels of vitamin E to incidence of sepsis and necrotizing entero 40 colitis in infants with birth weight 1500 grams or less.Pediatrics 75:619-638.

Johnson L, Schaffer D, Boggs TR. 1974. The premature infant, vitamin E deficiency, and retrolental fibro#-plasia. Am J Clin Nutr 27:1158-73.

Johnson L, Schaffer D, Quinn G, et al., 1982. Vitamin E supplementation in retinopathy of prematurity.Ann NY Acad Sci 393:473-95.

Keith CG, Kitchen WH. 1984. Retinopathy of prematurity in extremely low birthweight infants. Med J Aust 141:225-27.33

Kinsey VE, Jacobus JT, Hemphill FM. 1956. Retrolental fibroplasia: cooperative study of retrolental fibroplasia and the use of oxygen. Arch Ophthalmol 56:481.

Kinsey VE, Arnold HJ, Kalina RE, et al., 1977. PaO2 levels and retrolental fibroplasia: a report of the cooperative study. Pediatrics 60:655-668.

Knight NE, Roberts RJ. 1985. Tissue vitamin E levels in newborn rabbits after pharmacologic dosing. Develop Pharmacol Therap 8:96-106.

Kretzer FL, Hittner HM, Johnson AT, et al., 1982. Vitamin E and retrolental fibroplasia: ultrastructural support of clinical efficacy. Ann NY Acad Sci 393:145-166.

Laupus WE, Bousquet FP Jr. 1951. Retrolental fibroplasia: the role of hemorrhage in its pathogenesis. Am JD is Child 81:617-26.

Laurenti FR, Ferro R, Marzetti G, et al., 1980. Neutrophil chemotaxis in preterm infants with infections. J Pediatr 96:468-70.

Lorch V, Murphy D, Hoersten LR, et al., 1985. Unusual syndrome in premature infants: Association with a new intravenous vitamin E product. Pediatrics 75:598-602.41

Lucy JF, Dangman B. 1984. A reexamination of the role of oxygen in retrolental fibroplasia. Pediatrics 73:82-96.

McClung J. Backes C, Lavin A, et al., 1980. Prospective evaluation of vitamin E therapy in premature infants with hyaline membrane disease (HMD). Pediatr Res 14:604.

McCormick AQ. 1977. Retinopathy of prematurity. Cur Prob in Pediatr 7:1-28.

McPherson AR, Hittner HM. 1979. Scleral buckling in 2-1/2 to 11 month old premature infants with retinal detachment associated with acute retrolental fibroplasia. Ophthalmology 86:819-35.

Milner RA, Watts JL, Paes B, et al., 1981. RLF in 1500 gm neonates: part of a randomized clinical trial of the effectiveness of vitamin E. Retinopathy of Prematurity Conference, Washington, D.C., 2:703-716.

Nelson JS. 1983. Neuropathological studies of chronic vitamin E deficiency in mammals including humans. Biology of vitamin E. Pitman Books, London (Ciba Foundation symposium 101), p92-105.

Northway WH, Jr. 1978. Bronchopulmonary dysplasia in vitamin E. N Engl J Med 299:599-601.

Okuma M, Takayama H: 1980. Generation of prostacyclin and lipid peroxidation in vitamin E-deficient rats.Prostaglandins 19:527-536.39

Oski FA, Barness LA. 1967. vitamin E deficiency: a previously unrecognized cause of hemolytic anemia in premature infants. J Pediatr 70:211-220.

Oski FA. 1977. Metabolism and physiologic roles of vitamin E. Hospital Practice 79-85.

Owens WC, Owens EU. 1949. Retrolental fibroplasia in premature infants. Am J Opthalmol 32:1-21.

Owens WC, Owens EU. 1949. Retrolental fibroplasia in premature infants: II. Studies on the prophylaxis of the disease; The use of alpha tocopheryl acetate. Am J Opthalmol 32:1631-1637.

Patz A, Hoeck LE, De La Cruz E. 1952. Studies on the effect of high oxygen administration in retrolental fibroplasia: nursery observations. Am J Ophthalmol 35:1248-52.

Phelps DL. 1981. Local and systemic reactions to the parenteral administration of vitamin E. Develop Pharmacol Therap 2:156-171.

Phelps DL, Rosenbaum AL. 1977. The role of tocopherol in oxygen-induced retinopathy: kitten model. Pediatrics 59 (suppl):998-1005.

Phelps DL, Rosenbaum AL. 1979. Vitamin E in kitten oxygen-induced retinopathy. II. Blockage of vitreal neovascularization. Arch Ophthalmol 97:1522-26.

Phelps DL. 1981. Lack of drug effect on oxygen-induced retinal artery constriction in the cat. Pediatr Res 19:160A.

Phelps DL. 1981. Retinopathy of prematurity: an estimate of vision loss in the United States. Pediatrics 67:924-26.

Phelps DL. 1982. Vitamin E and retrolental fibroplasia in 1982. Pediatrics 70:420-425.

Phelps DL, Rosenbaum A, Isenberg SJ, et al., 1985. Effective IV tocopherol (Vit E) on retinopathy of prematurity (ROP). Pediatr Res 19:357A.

Phelps EL, Rosenbaum AL. 1979. Observations of vitamin E in experimental oxygen-induced retinopathy. Ophthalmology 86:1741-48.

Procianoy RS, Garcia-Prats JA, Hittner HM, etal., 1981. An association between retinopathy of prematurity and intraventricular hemorrhage in very low birthweight infants. Acta Pediatr Scand 70:473-477.

Pucklin JE, Simon RM, Ehrenkranz RA. 1982. Influence on retrolental fibroplasia of intramuscular vitamin E administration during respiratory distress syndrome. Opthalmology 89:96-102.

Purohit DM, Ellison RC, Zierler S, et al., 1985. Risk factors for retrolental fibroplasia: experience in 3025 premature infants. Pediatrics 76:339-344.

Reese AB, Blodi FC. 1951. Retrolental fibroplasia. Am J Ophthalmol 31:1-24.

Reisner SH, Amir J, Shohat M, et al., 1985. Retinopathy of prematurity: incidence and treatment. Arch Dis Child 60:698-701.

Roberts RJ, Knight, Mortensen ML, et al., 1985. Vitamin E content of tissues obtained from human infants given pharmacologic doses of tocopherol or tocopherol acetate intravenously. Pediatr Res 19:178A.

Rosenbaum AL, Phelps DL, Isenberg SJ, et al., 1985. Retinal hemorrhage in retinopathy of prematurity associated with tocopherol treatment. Opthalmology 92:1012-1014.38

Rosenblum JL, Keating JP, Prensky AL, et al., 1981. Progressive neurologic syndrome in children with chronic liver disease. N Engl J Med 304:503-8.

Saldanha RL, Cepeda EE, Poland RL. 1982. The effect of vitamin E prophylaxis on the incidence and severity of bronchopulmonary dysplasia. J Pediatr 101:89-93.

Schaffer DB, Johnson L, Quinn GE, et al., 1985. Vitamin E and retinopathy of prematurity: Follow-up at one year. Opthalmology 92:1005-1011.

Shigeoka AO, Santos JI, Hill H. 1978. Functional analysis of neutrophil granulocytes from healthy infected and stressed neonates. J Pediatr 95:454-59.

Silverman WA. 1980. Retrolental fibroplasia: A modern parable. Grune and Stratton, New York, NY, p2.

Smith IJ, Buchanan IG, Goss I, et al., 1969. Vitamin E and retrolental fibroplasia. N Engl J Med 309:369.The Committee for the Classification of Retinopathy of Prematurity. 1984. An international classification of retinopathy of prematurity. Arch Ophthalmol 102:1130-4.

Speer ME, Blifield C, Rudolph AJ,et al., 1984. Intraventricular
    hemorrhage and vitamin E in the very low birthweight infant:
    evidence for efficacy of early intramuscular vitamin E
    administration. Pediatrics 74:1107-1112.
Stuart MJ, Graeber JE, Clark DA. 1981. Neonatal vascular
    prostaglandin I2 and plasma prostaglandin I2 regenerating
    activity. Retionopathy of Prematurity Conference 1:194-202.
Terry TL. 1942. Extreme prematurity and fibroblastic overgrowth of
    persistent vascular sheath behind each crystalline lens. I.
    Preliminary report. Am J Ophthalmol 25:203-204.
Terry TL. 1945. Retrolental fibroplasia in the premature infant:
    further studies on fibroblastic overgrowth of persistent
    tunicavasculosa lentis. Arch Ophthalmol 33:203.
Watts JO, Milner RA, McCormick AQ. 1985. Failure of vitamin E to
    prevent RLF. Clin Invest Med 8:A176.
Wender DF, Thulin GE, Walker Smith GJ,et al., 1981. Vitamin E
    affects lung biochemical and morphologic response to hyperoxia
    in the newborn rabbit. Pediatr Res 15:262-268.
Wright W, Ank BJ,Herbert J, et al., 1975. Decreased bactericidal
    activity of leukocytes of stressed newborn infants. Pediatrics
    56:579-84.

Epidemiology of Age-Related Macular Degeneration

Leslie Hyman

I. REVIEW OF EPIDEMIOLOGIC METHODS

The primary goal of most epidemiologic research is to identify etiologic or risk factors for disease. Unlike most laboratory research in which an investigator can usually control experimental conditions, epidemiologic research is often limited to observational studies. Therefore, in these studies, epidemiologists observe and compare characteristics among different groups of people but, other than in clinical trials, cannot control which individuals have a particular exposure and which do not. However, these observations of groups of individuals are important to understanding the factors related to developing disease, since the exposures are observed under the "natural laboratory" or real life conditions.

Epidemiology, by definition (Last, 1983), is the study of the distribution and determinants of disease in populations. To determine the distribution of a particular disease in a population, first, a descriptive profile of those individuals who have the disease is developed according to the characteristics of person, place, and time, using available incidence and prevalence data. This profile includes data such as how frequently the disease occurs among different age, race, and sex categories, different countries or cities of residence, and whether the disease incidence has changed over time. These data are useful in providing initial leads towards identifying possible risk factors.

After the descriptive data have been developed, analytic studies are planned based on the information provided by these data, clinical observations, and experimental evidence to explore the associations with the suggested risk factors. There are two major types of analytic studies—case-control and cohort studies. The type of analytic studies most frequently conducted when beginning to investigate risk factors for a particular disease (e.g., age related macular degeneration (AMD), about which there is limited etiologic information, are the case-control studies. These studies, which are also called retrospective or case-referent studies, involve: 1) selecting a group of individuals with the disease (cases) and a comparable group of individuals without the disease (controls); 2) determining past exposure to selected potential risk factors in the case and the control groups; and 3) comparing the frequency of exposure to these potential risk factors in both study groups.

The advantages of case-control studies are that they are relatively inexpensive, can take a relatively short time to carry out if cases and controls are readily available, and are especially useful in studying rare diseases. They can also be used to evaluate simultaneously many causal hypotheses, whether they have been previously evaluated or are new, and therefore, occasionally are referred to as "fishing expeditions". The results of case-control studies can only estimate the risk of disease and cannot measure this risk directly. However, these studies are useful in generating hypotheses for larger prospective or cohort studies which directly measure the risk of disease (Lilienfeld and Lilienfeld, 1980). Thus, the numerous advantages of case-control studies make them an increasingly important tool of epidemiologic research. Several case-control studies have been conducted investigating AMD risk factors and have provided useful etiologic leads. They will be discussed in a later section.

II. AGE RELATED MACULAR DEGENERATION

A. Clinical Background

Age related macular degeneration (AMD)* is a leading cause of visual loss in the United States and has been found to occur in 8.8% of adults 52 years and older in one U.S. study (Kahn and Leibowitz, 1977). It has been reported by the National Society to Prevent Blindness as the leading cause of new cases of legal blindness among older individuals (National Society to Prevent Blindness, 1980). As increasing numbers of individuals reach these older ages of at least

---

* AMD is commonly called senile macular degeneration. However, the name has been changed by some ophthalmologists to make it less offensive to patients, since the term "senile" reflects only an age category and not the patient's status.

65 years, the social and economic consequences of blindness from AMD
will also increase unless successful means of prevention and
treatment are found. Unlike other major causes of blindness, such
as cataracts, in which a large proportion of patients are treatable,
no treatment is available for most AMD patients. The encouraging
results of the Macula Photocoagulation Study, determined that argon
laser photocoagulation is effective in preventing severe visual loss
from neovascularization associated with neovascular/exudative AMD
(Macula Photocoagulation Study Group, 1982). These results, while
demonstrating a significant breakthrough in treatment, are only
applicable to a small proportion of AMD patients who have the
neovascular form of AMD. Despite the magnitude of AMD as both a
clinical and a public health problem, its pathogenesis is only
partly understood, and its etiology remains unclear.

There are two major types of AMD – neovascular/exudative (N/E)
and atrophic (non-exudative). Each type is thought to be associated
with drusen, which are colloid bodies consisting of collections of
extracellular material located on or between Bruchs' membrane and
the retinal epithelium. They are believed to be an ophthalmoscopic
hallmark of AMD, but the precise nature of the relationship between
drusen and each AMD type has yet to be determined. All AMD patients
are believed to begin with drusen. As the disease progresses, zones
of apparent atrophy of the retinal pigment epithelium may occur in
the posterior pole of the eye. These changes can be associated with
a profound loss of central vision. This process constitutes the
atrophic form of the disease. An alternative course to this pigment
epithelial atrophy is the development of an elevated disciform
lesion with subretinal exudate and/or hemorrhage. Following this
course, formation of new blood vessels or neovascularization may
also be visible under the retina. Those individuals with subretinal
neovascularization may go on to develop disciform scars which lead
to incapacity and severe visual loss from AMD (Green and Key, 1977).
Although an estimated 80% of AMD patients have the atrophic form,
the N/E form may be responsible for almost 90% of the severe visual
loss (20/200 or worse) due to AMD (Ferris, et al., 1984).

B. EPIDEMIOLOGICAL BACKGROUND

Knowledge of the epidemiology of AMD is limited. This is due
in part to the relatively few epidemiologic studies on AMD that have
been reported. Many of these studies have been comprehensively

reviewed by Ferris (Ferris, 1983) and will also be discussed in this article. There are several sources of descriptive epidemiologic data on AMD. The most reliable sources are the Framingham Eye Study and the National Health and Nutrition Examination Survey (N-HANES) (Kahn and Liebowitz, 1977; Ganley and Roberts, 1983). Blindness registries are another source of data that have been used to estimate the prevalence of AMD when the condition is severe enough to cause legal blindness. Registries have been used for this purpose in England and Wales (Sorsby, 1966), Canada (MacDonald, 1965), and the United States (Kahn and Moorhead, 1973) and have identified AMD as a leading cause of blindness. The U.S. data is based on the Model Reporting Area study which was conducted using blindness registries from 16 states in the United States. The proportions of individuals registered as blind for the four leading causes of blindness in the three registries are compared in Table 1

Although registry data are suggestive of leading causes of blindness, their ability to estimate blindness prevalence or incidence is limited because of two serious problems. First, these data include only those persons who register as legally blind, and the proportion of those who are legally blind that register is not known. In fact, it is possible that as few as half of those who are actually legally blind register (Kahn and Moorhead, 1973; Graham, et al, 1968). A second potential problem is the lack of standardization in each country for classifying the cause of blindness. These limitations must be considered when interpreting the large differences in proportions of blind persons with AMD among the three registries presented in Table 1 and probably account in large part for these differences. However, some of this variation may still be due to true differences in these populations.

The N-HANES and Framingham Eye Studies are both population based studies that include data on the prevalence and descriptive epidemiologic features of AMD. The N-HANES survey was conducted between April 1971 and June 1974 and involved standardized examinations of persons selected from probability samples of non-institutionalized United States populations. Interpretation of the results of this study may be limited by the fact that only 70% of the selected population received eye exams and the examiners had diverse levels of experience and training (Ganley and Roberts, 1983). These possible limitations of the data should be kept in mind when interpreting Table 2. The data presented in this table

TABLE 1

PROPORTION OF REGISTERED BLIND FROM THE
FOUR LEADING CAUSES OF BLINDNESS

Cause	England & Wales Registered 1955-1962	United States Registered 1970	Canada Living Registrants 1964
AMD	26%	13%	5%
Cataract	23%	12%	15%
Glaucoma	12%	11%	10%
Diabetic Retinopathy	7%	11%	5%
Total Registered Blind	60,309	8,353	24,605

(Ferris, 1983)

suggests an increase in AMD prevalence with age (Klein and Klein, 1982). AMD has been observed in the Framingham Eye Study and in other clinical studies to be more prevalent in females than males. It has also been observed to be more frequent in whites than blacks, but these observations are mostly based on clinical case series. The findings of the N-HANES survey are not consistent with these data. In fact, among whites AMD appeared slightly less prevalent in females (Klein and Klein, 1982). However, these data must be interpreted cautiously because of their limitations.

The Framingham Eye Study, conducted by Kahn et al between 1973-1975, involved performing comprehensive standardized ophthalmological examinations on 2631 persons aged 52-85 who were residents of Framingham, Massachusetts. This study group comprised 2/3 of the surviving members of the cohort who had participated in the Framingham Heart Study. The study used specific diagnostic criteria for AMD which included pigment disturbance in the macula, drusen, perimacular circinate exudates, and serous, hemorrhagic or proliferative elevation of the pigment epithelium. The examining ophthalmologist classified an eye as having "senile macular degeneration" if the etiology of the specific lesions was determined as "senile" and the best corrected visual acuity was 20/30 or worse (Kahn, et al., 1977; Leibowitz, et al., 1980). The visual acuity criteria did not specify that the decrease in acuity must be due to macular disease and thus created a problem of disease definition in this study. As a result, eyes were probably categorized as AMD when some pigmentary change was present, but the visual loss was due, in

TABLE 2

NHANES-I PREVALENCE RATES OF
SENILE MACULAR DEGENERATION
BY AGE, RACE, AND SEX

Sex and Race	Age 45-64 Yrs.		Age 65-75 Yrs.	
	N	%	N	%
White				
Male	513	2.3	612	9.6
Female	561	2.3	654	6.9
Black				
Male	156	3.8	193	9.3
Female	165	2.4	184	11.4

(Ferris, 1983)

fact, to another reason, such as senile cataract. Conversely, eyes were probably classified as having senile cataract when the visual loss was caused by AMD. Despite this potential limitation, this study has provided the best available descriptive epidemiologic data on AMD. These data demonstrate the local population prevalence of AMD for people over the age of 52 to be 8.8% (Table 3). They also demonstrate that AMD prevalence increases with age and remains higher for women than men in each age group (Table 3) (Kahn, et al., 1977). The observation that AMD is higher for females differs from the N-HANES data.

A summary of the epidemiologic characteristics of AMD reflects our limited knowledge of the disease (Table 4). The only factor found to be associated with AMD in all studies is that of age. The prevalence of the disease increases with age, cases are usually 50 years and older, and there is a rapid increase in prevalence among persons 65 years and older.

Clinical studies have suggested along with clinical impressions that AMD is more common among whites than blacks (Gregor and Joffe, 1978; Chumbley, 1977), but this was not found to be true in the N-HANES survey. With respect to sex, the Framingham Eye Study found AMD to be approximately 50% more prevalent in women than in men in each of the age groups studied. However, Sperduto and Seigel found no increased prevalence of mild or more severe senile macular changes among females compared to males when they carried out additional analyses of the Framingham data. These data were based only on the presence of the lesions of AMD seen at the eye examination and not on a decreased visual acuity criteria included

TABLE 3

PREVALENCE OF AMD BY AGE AND SEX: FRAMINGHAM EYE STUDY (1973-1975), LOCAL AREA ONLY

Age (years) and sex	Total Persons Screened	% With AMD
Total	2477	8.8
Men	1043	6.7
Women	1434	10.3
Age 52-64	1293	1.6
Men	573	1.2
Women	720	2.0
Age 65-74	787	11.0
Men	318	8.8
Women	469	12.6
Age 75-85	397	27.9
Men	152	24.4
Women	245	30.1

(Ferris, 1983)

for the prior analyses (Sperduto and Seigal, 1980). The analysis of the N-HANES-1 data by Klein and Klein shows no increase in females compared to males with either senile macular degeneration or the presence of drusen or retinal pigment epithelial changes (Table 2) (Klein and Klein, 1982).

The data from the N-HANES survey, therefore, has raised doubts both about the impression that males are at lower risk for developing AMD than females and the impression that blacks are at lower risk of developing this disease than whites.

C. RISK FACTORS

Table 5 summarizes a number of reported risk factors other than age, race, and sex which have been suggested as being associated with AMD by certain epidemiologic and clinical studies. The Framingham Eye Study is the best population based source of information on possible risk factors for AMD. This study used the age-sex specific data previously collected in the Framingham Heart Study, together with the ophthalmic diagnoses made in the Framingham Eye Study, to positively associate the variables of elevated systemic blood pressure, short height, decreased vital capacity, left ventricular hypertrophy, decreased hand grip strength, and history of lung infection with AMD (Kahn, et al., 1977). Since a

TABLE 4

SUMMARY OF AMD
EPIDEMIOLOGIC CHARACTERISTICS

AGE: Prevalence increases with age
          Cases usually 50 years and older
RACE: May be more prevalent in whites than blacks
SEX: May be more prevalent in females than males

large number of variables was investigated, it is likely that these findings are a mix of real and chance associations which need to be corroborated.

There are two case-control studies included in Table 5, one by Delaney and Oates (Delaney and Oates, 1982) and the other by Maltzman, Mulvihill and Greenbaum (Maltzman, et al., 1979). Both of these studies found associations between hyperopia and AMD; Delaney and Oates also found an association between hypertension and AMD. These studies are limited by their small sample sizes of 50 and 30 age and sex matched cases and controls, respectively, and problems with control group selection. However, the consistency of the findings between these two studies regarding hyperopia and between the Delaney and Oates study and the Framingham Eye Study regarding hypertension lends some credibility towards these associations. Case reports and clinical studies have also associated macular degeneration with family history of macular disease (Gass, 1973) and cigarette smoking, (Paetkau, et al., 1978), as well as hypertension.

Hyman et al has conducted the most extensive case-control study on AMD risk factors to date. The study has been described in detail elsewhere and will be summarized in this discussion (Hyman, et al., 1983). The population for this study included individuals who visited any of 34 Baltimore ophthalmologists between September 1978 and March 1980 and who met the study criteria for cases and controls. The selection criteria for the cases were:

1) seen since September 1, 1978 with a diagnosis of drusen and/or macular degeneration by a participating ophthalmologist;

2) experienced some vision loss that the ophthalmologist thought was due to macular degeneration and not to other ophthalmic disease;

3) age less than 85 years at the time of identification;

4) both pupils could be dilated;

5) media were clear enough for evaluating the fundus from photographs;

TABLE 5

REPORTED ASSOCIATIONS OF RISK FACTORS AND AMD

Risk Factor	Framingham Eye Study[*]	Delaney & Oates[**]	Maltzman et al[+]	Other Clinical Studies[++]
Elevated Systemic Blood Pressure	+	+	-	+
Decreased Handgrip Strength	+	O	O	-
Deceased Vital Capacity	+	O	O	O
Left Ventricular Hypertrophy	+	O	O	O
Short Height	+	O	O	O
History of Lung Infection	+	O	O	O
Cigarette Smoking	-	O	-	+
Hyperopia	O	+	+	O
Family History of AMD	O	-	O	+

+   Positive Association

-   No Association

O   Not Studied

```
 *  Kahn, et al., 1977
**  Delaney and Oates, 1982
 +  Maltzman, et al., 1979
++  Gass, 1973; Paetkau, et al., 1978; Vidaurri, et al., 1984
```

6) resided in the Standard Metropolitan Statistical Area (SMSA) and had a telephone; and

7) mobile enough to come to a participating clinic or hospital for the study appointment (Hyman, et al., 1983).

The selection criteria for controls were:

1) age within five years of the age of the case (in addition, cases 70 years and older were matched to controls 70 years and older, and cases less than 70 years were matched to controls less than 70 years); and

2) sex (Hyman, et al., 1983).

The controls were individuals seen for routine eye examinations, had no severe eye pathology, and were similar to the cases with respect to residence, media clarity, ability for pupil dilation, and mobility. Eighteen percent of the controls did have mild cataracts which is not unexpected among individuals in this age category. However, the percentage of cases with mild cataracts was similar.

Two hundred and twenty-eight AMD cases and 237 controls matched on age-sex and ophthalmologist were interviewed at one of three hospitals in Baltimore for past medical, residential, occupational, smoking, and family histories, as well as social and demographic factors. Measurements included blood pressure, visual acuity, height, hand grip strength, refractive error, and determination of iris color. Stereo fundus photographs were taken to confirm the presence or absence of macular degeneration. Further information on family history was ascertained by questionnaire to family members and to their eye doctors. Medical histories were validated by questionnaires to the physicians of the study participants. Of the 228 AMD cases and 237 controls interviewed, 162 AMD cases and 175 controls had fundus photos that were consistent with the referring diagnosis. This group of 162 cases and 175 controls was used for the final data analyses.

A large number of factors were screened for possible associations with AMD, many of which on initial comparison between AMD cases and controls demonstrated no association between the specific factor and the disease. These factors included parental birthplace, occupational history, history of non-cardiovascular disease including cancer, tuberculosis, stomach or intestinal problems, diabetes, blood problems, medication history, radiation history, height, sunlight exposure, history of cigar and pipe smoking, and family medical and ophthalmic histories other than AMD.

The positive findings were grouped according to the three categories of familial history of AMD, specifically, maternal and sibling AMD histories, environmental factors of non-specific chemical work exposures and cigarette smoking in males only, personal characteristics of blue or medium pigmented eyes, history of cardiovascular diseases (including heart disease, angina, other heart problems, and stroke), decreased hand grip strength, and hyperopia.

Family history of AMD was evaluated by combining the parents and siblings of the study participants (Table 6). For this analysis family history of AMD was considered to be positive if either a parent or a sibling of an AMD case or a control had a reported history of the disease. Information on parental AMD history was based on responses from the study participants; information on

TABLE 6

FREQUENCY OF HISTORY OF AMD AMONG FAMILY MEMBERS
OF AMD CASES AND CONTROLS

	Cases	Controls	Odds Ratios
	(% With Family AMD History)		(95% Confidence Limits)
Positive Family*			
AMD History	21.6 (162)	8.6 (175)	2.9
			(1.5-5.5)

* Positive family AMD history is defined by a reported AMD history
in at least one family member (parent or sibling).

(Hyman, et al., 1983)

sibling AMD history was based on responses from their eye care
providers. Table 6 presents the percent of AMD cases and controls
with at least one family member with AMD and demonstrates that 21.6%
of the AMD cases and 8.6% of the controls had at least one family
member with AMD. The prevalence of AMD among control family members
of 8.6% is similar to the Framingham AMD prevalence of 8.8% and
therefore gives more credibility to this finding. The comparison
between the AMD cases and controls results in an odds ratio of
approximately 3. This suggests a 3-fold increased risk of developing macular degeneration if either a parent or a brother or sister
also has the disease.

The environmental factors presented in Table 7 were found to be
associated with AMD. The association with chemical work exposure,
which is demonstrated by an odds ratio of 4.2, was based on the
response to the question, "Did you ever work around chemicals which
caused your eyes to burn, on a regular basis?" This question was
one of a series of questions on a variety of occupational exposures.
Responses included a variety of chemicals with varying durations of
exposure. This association was stronger for males than females with
10 out of 11 cases reporting this exposure being males and only one
being female. The association between AMD and cigarette smoking was
found in males only. The difference in the significance of these
associations between males and females may be due to the minimal
exposure of the women in the older age groups to these factors.

The association between AMD and iris color can be viewed as
either a positive association between AMD and blue or medium
pigmented eyes or a negative or protective association with brown
eyes (Table 8). This association was statistically significantly

TABLE 7

ENVIRONMENTAL FACTORS OF NONSPECIFIC CHEMICAL WORK EXPOSURES
AND CIGARETTE SMOKING AMONG AMD AND CONTROLS

Exposure History	AMD Cases	Controls	Odds Ratio
	(Percent With Exposures)		(95% Confidence Limits)
Nonspecific chemical work exposures (both sexes)	6.2 (162)	1.7 (175)	4.2 (1.1-15.2)
Cigarette Smoking (males only)	83.1 (65)	65.2 (66)	2.6 (1.1-5.7)

(Hyman, et al., 1983)

stronger for males than females.

The analysis of refractive error and hand grip strength was done using a t-test of the means (Table 9). Refractive error was significantly associated with AMD when both sexes were analyzed together, but when each sex was analyzed separately it was positively associated in females (p = .01) but not in males, with the female cases having a more positive refractive error or being more hyperopic than the female controls. This analysis has excluded patients who are aphakic, since they have had an artificially high positive refractive error induced by the surgery (Hyman, et al, 1983).

Hand grip strength was found to be significantly associated in both males and females in this study, with the male and female cases having weaker hand grip strength than the male and female controls (Hyman, et al., 1983).

Independent statistically significant associations were found between AMD and a history of stroke, transient ischemic attack, circulatory problems, and arteriosclerosis. The associations of a history of myocardial infarction and other heart problems, although not statistically significant, were in a positive direction. There was no association found between AMD and hypertension. Since a number of cardiovascular problems were positively associated with AMD with varying levels of statistical significance and these conditions are related to each other, they were combined for further analysis. The association between AMD and a history of one or more cardiovascular diseases, presented in Table 10, includes the conditions of myocardial infarction, angina, other heart problems, stroke or transient ischemic attack, arteriosclerosis, circulatory

TABLE 8

IRIS COLOR AMONG AMD CASES AND CONTROLS

Iris Color	AMD Cases (Percent With Exposures)		Controls		Odds Ratio (95% Confidence Limits)
Blue (light)	41.4		33.9		3.5 (1.7-6.6)
Medium pigment	49.4	(162)	39.7	(174)	3.6 (1.6-8.4)
Brown (dark)	9.2		26.4		1.0
Males Only					
Blue or medium pigment	95.4	(65)	71.2	(66)	8.3 (2.3-29.7)
Female Only					
Blue or medium pigment	87.6	(97)	75.0	(108)	2.4 (1.1-5.0)

(Hyman, et al., 1983)

problems, and hypertension. The results are at borderline levels of statistical significance with an odds ratio of 1.7 (Hyman, et al, 1983).

In addition to these numerous case/control comparisons, we also compared the two major types of AMD, neovascular/exudative and atrophic, with respect to a variety of factors. Neovascular/exudative AMD was found to be positively associated with older ages, visual acuity of 20/200 or worse, a history of cardiovascular disease, decreased hand grip strength, and hyperopia (Table 11) (Hyman, 1981).

There have also been two other case-control studies conducted recently by Weiter et al. One study of 650 white AMD patients and 363 controls investigated the association of AMD with ocular (iris and fundus) pigmentation (Weiter, et al., 1985); the second study of 49 AMD patients and 27 age-matched controls investigated the role of selenium nutrition in AMD by using glutathione peroxidase activity as an indicator of the adequacy of selenium nutritional status (Weiter, et al., 1985). The results of the first study corroborated the findings of Hyman et al by indicating an association between lightly pigmented irises; the second study suggested an inverse correlation between AMD severity and the level of glutathione peroxidase in the plasma. There are possible biases in both of these studies in the selection of the cases and controls. Also, the sample size of the second study is small and the results are preliminary. However, the consistency of the findings regarding ocular pigmentation with Hyman et al makes this association more likely. The association with glutathione peroxidase activity and

TABLE 9

COMPARISON OF REFRACTIVE ERROR AND HAND GRIP STRENGTH
AMONG AMD CASES AND CONTROLS

	AMD Case	Controls	p
Males			
Mean refractive error*	1.4	.9	.16
Mean hand grip strength	32.2	36.3	.01
Female			
Mean refractive error*	1.8	1.1	.01
Mean hand grip strength	18.5	20.4	.02

* excluding aphakics

(Hyman, et al., 1983)

selenium, although suggestive at best, should be pursued further,
since it is consistent with animal and laboratory hypotheses
regarding antioxidants (Hayes, 1974; Katz, et al., 1978; Katz, et
al., 1982; Stone, Dratz, 1982), ocular pigmentation (Lavail, 1980),
and certain types of retinal degenerations.

Summary and Conclusion

The risk of AMD appears to be based on a combination of
familial characteristics, environmental factors, personal
characteristics, and associated diseases. Available data suggest
that AMD 1) risk increases with increasing age; 2) may be more
prevalent among females than males; and 3) may be more prevalent
among whites than blacks. Factors which have consistently been
found by several epidemiologic studies to be associated with AMD are
increased blood pressure and/or history of cardiovascular disease,
light iris color, family history of AMD, hyperopia, and hand grip
strength.

Although evidence for most of the AMD risk factors studied is
still only suggestive, there is some consistency among different
studies regarding associations between AMD and hypertension, other
cardiovascular diseases, ocular pigmentation, hyperopia, and hand
grip strength.

The Framingham Eye Study (FES), as previously mentioned, found
AMD to be associated with high diastolic and systolic blood pressure
and hypertension history (Kahn, et al., 1977). Further analyses of
these data by Sperduto and Hiller also demonstrated a progressive
increase in the odds ratios, or degree of association, with
increased duration of hypertension (Sperduto, Hiller, 1985). Thus,

TABLE 10

HISTORY OF ONE OR MORE CARDIOVASCULAR DISEASES AMONG AMD CASES
AND CONTROLS

	AMD Cases	Controls	Odds Ratio
	(% With Positive History)		(95% Confidence Limits)
History of one or more cardiovascular diseases*	74.7 (162)	63.4 (175)	1.7 (1.1-2.7)

* History of cardiovascular diseases includes myocardial infarction, angina, other heart problems, arteriosclerosis, hypertension, other circulatory problems, stroke and/or transient ischemic attacks.

(Hyman, et al., 1983)

hypertension at an early age may increase the risk of AMD. Using a similar approach of correlating eye examination data with previously collected data on blood pressure, an Israeli study by Vidaurri, et al also showed an association between drusen and elevated blood pressure (Vidaurri, et al., 1984). While the interpretation of this finding may be limited by small numbers and problems of investigating numerous factors simultaneously, it does corroborate the results of the FES. In the National Health and Nutrition Examination Survey (N-HANES), mean systolic blood pressure was slightly higher for macular degeneration cases than noncases. Although the statistical significance of this association was not consistent for all age-sex-race categories, this could be due to the small number of cases in some categories. While the Delaney and Oates study of 50 cases and their age-sex matched controls found an association between AMD and antihypertensive drug use, the study by Maltzman et al of 30 cases and their matched controls found no association between AMD and vascular disease. However, both studies were small and have possible biases in control group selection. Hyman et al found no association of AMD with blood pressure, hypertension history, or use of antihypertensives. However, they did find AMD to be associated with arteriosclerosis, circulatory problems, and stroke and/or transient ischemic attacks; a history of myocardial infarction and other cardiac problems was also found more frequently in AMD cases than controls. They also noted that a history of cardiovascular disease was associated with N/E AMD. Their inability to identify hypertension as a risk factor in this study may be due in part to the high prevalence of hypertension in

TABLE 11

FREQUENCY OF CHARACTERISTICS POSITIVELY ASSOCIATED WITH NEOVASCULAR/EXUDATIVE MACULAR DEGENERATION

| Characteristic | Percent With Specified characteristic among cases with | | | Odds Ratio (95% Confidence Limits) |
	Neovascular/ Exudative Cases (N)	Atrophic Cases (N)	p*	
Visual acuity of 20/200 or less	83.5 (79)	12.7 (71)	<.001	35.0 (13.4-100.4)
History of cardiovascular diseases[+]	81.0 (79)	64.8 (71)	.05	2.3 (1.1-4.8)
Low hand grip strength[#]	40.5 (79)	22.5 (71)	.03	2.3 (1.1-4.6)
Hyperopia§	60.3 (79)	45.6 (71)	.003	
Mean age $\pm$ SE (years)	72.8$\pm$.85	69.9$\pm$1.1	.04[@]	

* The p value is associated with the chi square test for independent samples.

[+] History of cardiovascular diseases includes myocardial infarction, angina, other heart problems, arteriosclerosis, hypertension, other circulatory problems, stroke and/or transient ischemic attacks.

[#] Hand grip strength was defined separately for males and females with low hand grip for males being less than 32 Kg and low hand grip for females being less than 15 Kg.

§ Spherical equivalent was defined according to three categories of myopia, emmetropia, and hyperopia. Myopes were defined as individuals with a measurement of spherical equivalent of less than -1; emmetropes were defined as individuals with spherical equivalent of -1 to +1, and hyperopes were defined as individuals with spherical equivalent of greater than +1. Individuals with one eye myopic and one eye hyperopic were excluded from the analysis. Aphakics were also excluded.

[@] The p value is associated with the results of a t-test for the comparison on means of independent observations.

(Hyman, et al.) Unpublished manuscript.

older age groups.

The association with light pigment color was initially suggested by the observation that AMD is less prevalent in blacks than whites. The finding by the Hyman study that AMD is less common in persons with brown eyes than in those with blue or medium pigmented eyes is consistent with this association. The case-control study by Weiter also suggests an association between light iris color and AMD. This observation raises the question of the nature of this association, or how iris pigmentation might influence the development and/or progression of AMD. Therefore, the role of pigmentation in general-skin as well as ocular-should be explored further.

Other factors suggested by epidemiologic and/or clinical studies which should be investigated further include prolonged exposure to sunlight, nutritional status with respect to certain nutrients such as selenium and vitamins E and C, occupational exposures to chemicals, skin pigmentation, and cigarette smoking.

Thus far, the factors that appear most likely to affect the development of AMD and type of AMD seem to be familial, genetic, and personal characteristics, rather than environmental factors. Epidemiologic studies with larger numbers of subjects are needed to evaluate further the role of environmental factors, such as dietary factors and sunlight exposure, in AMD. Although our knowledge of risk factors is limited, some good etiologic leads have been developed. Further studies are needed with adequate sample sizes to evaluate some of these possible risk factors. Since it is now known that a certain proportion of patients with neovascular/exudative AMD are eligible for treatment, it will be extremely useful to be able to target high risk groups. These groups, when identified, can be followed closely and treated at an early stage of disease so that visual loss can be delayed and possibly prevented. Other areas of research on this disease would involve identifying prognostic factors in order to prevent individuals with the disease from developing more serious complications.

REFERENCES

Chumbley, L. 1975. Impressions of eye diseases among Rhodesian blacks in Mashonaland. S. Afr. Med. J. 52:316-318.

Delaney, W. and R. Oates. 1982. Senile macular degeneration: A preliminary study. Ann. Ophthalmol. 14:21-24.

Ferris, F. 1983. Senile macular degeneration: Review of epidemiologic features. Am. J. Epidemiol. 118:132-151.

Ferris, F., S. Fine, L. Hyman. 1984. Age-related macular degeneration and blindness due to neovascular maculopathy. Arch. Ophthalmol. 102:1640-1642.

Ganley, J. and J. Roberts. 1983. Eye conditions and related need for medical care among persons 1-74 years of age, United States, 1971-72. Vital and Health Statistics, Series 11, no. 228, DHHS. Publication no. (PHS) 83-1678.

Gass, J. 1967. Pathogenesis of disciform detachment of the neuroepithelium. III. Senile disciform macular degeneration. Am. J. Ophthalmol. 63:617-644.

Gass, J. 1973. Drusen and disciform macular detachment and degeneration. Arch. Ophthalmol. 90:208-217.

Graham, P., J. Wallace, E. Walsby, et al. 1968. Evaluation of postal detection of registrable blindness. Br. J. Prev. Soc. Med. 22:238.

Green, W. and S. Key. 1977. Senile macular degeneration: A histopathologic study. Tr. Am. Ophth. Soc. 75:180-254.

Gregor, Z. and L. Joffe. 1978. Senile macular changes in the black African. Br. J. Ophthalmol. 62:547-550.

Hayes, K. 1974. Retinal degenerations in monkeys induced by deficiencies of vitamin E or A. Invest. Ophthalmol. 13:499-510.

Hyman, L. 1981. Senile macular degeneration: An epidemiologic case-control study. Thesis. The Johns Hopkins University, Baltimore, MD.

Hyman, L., A. Lilienfeld, F. Ferris, S. Fine. 1983. Senile macular degeneration: A case-control study. Am. J. Epidemiol. 118:213-227.

Hyman, L., F. Ferris, S. Fine, A. Lilienfeld. Senile macular degeneration: An epidemiologic investigation. Unpublished manuscript.

Kahn, H., H. Leibowitz, J. Ganley, et al. 1977. The Framingham Eye Study. I. Outline and major prevalence findings. Am. J. Epidemiol. 106:17-32.

Kahn, H., H. Leibowitz, J. Ganley, et al. 1977. The Framingham Eye Study II. Association of ophthalmic pathology with single variables previously measured in the Framingham Heart Study. Am. J. Epidemiol. 106:33-41.

Kahn, H. and H. Moorhead. Statistics on Blindness in the Model Reporting Area 1969-1970. Washington DC: USDHEW Publication No. (NIH) 73-427.

Katz, M., K. Parker, G. Handelman, et al. 1982. Effects of antioxidant nutrient deficiency on the retina and retinal pigment epithelium of albino rats; a light and electron microscopic study. Exper. Eye Res. 34:339-369.

Katz, M., W. Stone, E. Dratz. 1978. Fluorescent pigment accumulation in retinal pigment epithelium of antioxidant deficient rats. Invest. Ophthalmol. Vis. Sci. 17:1049-1058.

Klein, B. and R. Klein. 1982. Cataracts and macular degeneration in older Americans. Arch. Ophthalmol. 100:571-573.

Last, J. 1983. A Dictionery of Epidemiology. New York. Oxford University Press. pp. 32-33.

Lavail, M. 1980. Eye pigmentation and constant light damage in the rat retina. In The Effects of Constant Light on Visual Processes. T. William and B. Baker (eds.). pp. 357-387.

Leibowitz, H., D. Krueger, L. Maunder, et al. 1980. The Framingham Eye Study Monograph. Surv. Ophthalmol. 24 (Suppl):456.

Lilienfeld, A. and D. Lilienfeld. 1980. Foundations of Epidemiology. New York. Oxford University Press. pp. 191-218.

MacDonald, A. 1965. Causes of blindness in Canada. Can. Med. Assoc. J. 92:264-279.

Macula Photocoagulation Study Group. 1982. Argon laser photocoagulation for senile macular degeneration: Results of a randomized clinical trial. Arch. Ophthalmol. 100:912-918.

Maltzman, B., M. Mulvihill, A. Greenbaum. 1979. Senile macular degeneration and risk factors: a case-control study. Ann. Ophthalmol. 11:1197-1201.

National Society to Prevent Blindness. 1980. Vision Problems in the U.S. New York. National Society to Prevent Blindness.

Paetkau, M., T. Bogd, M. Grace, et al. 1978. Senile disciform macular degeneration and smoking. Can. J. Ophthalmol. 13:67-71.

Sarks, S. 1976. Aging and degeneration in the macular region: A clinicopathological study. Brit. J. Ophthalmol. 60:324-341.

Sorsby, A. 1966. Reports on Public Health and Medical Subjects. No. 114. London. Her Majesty's Stationery Office.

Sperduto, R. and R. Hiller. 1985. Personal communication.

Sperduto, R. and D. Seigal. 1980. Senile lens and senile macular changes in a population based sample. Am. J. Ophthalmol. 90:86-91.

Stone, W. and E. Dratz. 1982. Selenium and non-selenium glutathione peroxidase activities in selected ocular and non-ocular rat tissues. Exp. Eye Res. 35:405-412.

Vidaurri, J., J. Pe'er, S-T. Halfon, et al. 1984. Association between drusen and some of the risk factors for coronary artery disease. Ophthalmologica. 188:243-247.

Weiter, J., E. Dratz, K. Fitch, G. Handelman. 1985. Role of selenium nutrition in senile macular degeneration. Invest. Ophthalmol. Vis. Sci. Suppl. 26:58.

Weiter, J., F. Delori, G. Wing, K. Fitch. 1985. Relationship of senile macular degeneration to ocular pigmentation. Am. J. Ophthalmol. 99:185-187.

Index

DATE DUE

DATE DUE			
NOV 2 9 2000			
MAY 2 7 2005			

DEMCO NO. 38-298